Naveed Iqbal

Millimeter Wave Radio Channels: Properties, Multipath Modeling and Simulations

Millimeter Wave Radio Channels: Properties, Multipath Modeling and Simulations

Naveed Iqbal

Universitätsverlag Ilmenau
2023

Impressum

Bibliografische Information der Deutschen Nationalbibliothek
Die Deutsche Nationalbibliothek verzeichnet diese Publikation in der Deutschen Nationalbibliografie; detaillierte bibliografische Angaben sind im Internet über http://dnb.d-nb.de abrufbar.

Diese Arbeit hat der Fakultät für Elektrotechnik und Informationstechnik der Technischen Universität Ilmenau als Dissertation vorgelegen.

Tag der Einreichung:	25. Juni 2021
1. Gutachter:	Univ.-Prof. Dr.-Ing. habil. Reiner S. Thomä (Technische Universität Ilmenau)
2. Gutachter:	Univ.-Prof. Dr.-Ing. Christoph Mecklenbräuker (Technische Universität Wien)
3. Gutachter:	Associate Professor Dr. Katsuyuki Haneda (Aalto University Finland)
Tag der Verteidigung:	5. Dezember 2022

Technische Universität Ilmenau/Universitätsbibliothek
Universitätsverlag Ilmenau
Postfach 10 05 65
98684 Ilmenau
https://www.tu-ilmenau.de/universitaetsverlag

ISBN 978-3-86360-273-4 (Druckausgabe)
DOI 10.22032/dbt.56740
URN urn:nbn:de:gbv:ilm1-2022000589

Titelfoto: photocase.com | AlexFlint

I would like to dedicate this thesis to my loving family and teachers...

Acknowledgements

I did this work under the supervision of Prof. Reiner S. Thomä from Technische Universität Ilmenau as an industrial graduate student working at Huawei technologies (Munich research center) during the period May 2015 till April 2020. First of all, I would like to thank my advisor Prof. Reiner S. Thomä and Huawei for providing me this opportunity and exposure. I am grateful to my colleagues from the measurement team which include Robert Müller, Diego Dupleich, Stephan Häfner and Sergii Skoblikov for their tireless efforts during the measurement campaigns used in this work. I still remember, how these guys spent sleepless nights in carrying out these measurement campaigns particularly when they were allowed to use the site for a short duration. I highly appreciate efforts from my second supervisor Dr. Jian Luo and Christian Schneider (from TUI) for providing me timely feedback on the paper contents and providing suggestions that improved the overall quality of paper.

During this work, I also had some email exchanges with Dr. Waseem Malik from Massachusetts Institute of Technology, U.S.A. and Dr. Gulzaib Rafiq from University of Agder, Norway. I highly regard them for having useful discussions. I would also like to acknowledge Dr. Mate Boban and Richard A. Stirling-Gallacher for providing timely review on my papers.

I really appreciate the efforts from my colleague and friend Tobias Christian Lass who has written "*Zussamenfassung*" of this dissertation in Deutsch. Similarly, I would also like to thank my friends, Dr. Aftab Naqvi, Dr. Usman Sanwal, Dr. Sher Ali Cheema and Anastasios Kakkavas for doing much needed proof-reading of this dissertation for grammatical mistakes and sentence corrections.

Finally, I would like to highly regard the hospitality of Munich city. Although it is one of the most expensive cities of Germany, but I would admit that it was not so hard on me. It provided me a relatively low priced accommodation, health and child care to my family which made things quite favourable during Ph.D. study.

Abstract

Studies presented in this thesis are based on the characterization of realistic radio channels obtained from exhaustive channel sounding campaigns. Mainly, three fundamental problems of wireless channel modelling have been investigated for millimetre wave (mmWave) radio channel modelling application, namely (i) Frequency dependence of propagation, (ii) Impact of antenna directivity on the channel model definition, and (iii) Impact of system bandwidth on the radio channel modelling. A detailed description of these problems is as follows:

(i) **Frequency Dependence of Propagation.** Multi-band measurement campaigns are carried out using directional antennas which do an omni-directional scan of the propagation environment. During the measurements, Tx-Rx systems are placed at fixed positions and the propagation environment remained as static as possible. Using synthesized omni-directional power delay profiles (PDPs), we aim to investigate if there exists a frequency dependency in the multipath dispersion statistics, e.g. delay and angular spreads.

(ii) **Impact of Antenna Directivity on the Channel Model Definition.** Small-scale fading measurements are carried out which emulate a scenario, where a radio communication link is established through a single multipath cluster which is illuminated using antennas with different Half Power Beam Widths (HPBW). The major goal here is to investigate the impact of spatial multipath filtering on the small-scale fading due to high antenna directivity. In particular, the impact on variations in the receive signal strength and the validity of narrowband wide-sense stationary assumption (both in time and frequency domains) is investigated.

(iii) **Impact of System Bandwidth on the Radio Channel Modelling.** Small-scale fading measurements are used to illuminate multipath clusters in a lecture room scenario. The primary objective is to investigate the impact of high system bandwidth on variations in the receive signal strength, randomness in the cross-polarization power ratio (XPR) and richness of the multipath scattering.

Based on the characterization of realistic radio channels, results presented in this dissertation lead towards an understanding that when moving up to the higher frequencies, frequency itself does not play a significant role in defining the channel modelling methodology. In fact, how a propagation channel is illuminated is of fundamental importance. Therefore, mmWave system properties such as a high antenna directivity and system bandwidth are shown to have a high influence on the channel model definition. In general, fade depth scaling as a function of system bandwidth is quite well understood. We demonstrate that, the high antenna directivity of mmWave systems result in a further reduction in the fading depth. In addition, we explore some new directions to this line of research which are based on the second-order statistical analysis of the channel impulse response (CIR) vector. Our results emphasize that, fading statistics of resolvable channel taps in a mmWave radio channel cannot be modelled as Rayleigh-Rice distributed random variables. This is primarily due to the fact that channels with sparse scattering conditions are illuminated due to high antenna directivity and bandwidth of mmWave systems. Consequently, the complex Gaussian random variable assumption associated with Rayleigh-Rice fading distributions does not remain valid. Further, it has been demonstrated that, high antenna directivity and bandwidth of mmWave systems also raise a question mark on the validity of wide-sense stationary (WSS) assumption in the slow-time domain of mmWave radio channels. Results presented in this contribution are novel and they provide theoretically consistent insights into the measured radio channel.

Zusammenfassung

In dieser Arbeit werden drei grundlegende Probleme der Modellierung von Drahtloskanälen für die Anwendung bei der Funkkanalmodellierung im Millimeterwellenbereich (mmWave) untersucht, nämlich (i) die Frequenzabhängigkeit der Ausbreitung, (ii) der Einfluss der Antennenrichtwirkung auf die Definition des Kanalmodells und (iii) der Einfluss der Systembandbreite auf die Funkkanalmodellierung. Die detaillierte Beschreibung dieser Probleme lautet wie folgt:

(i) **Frequenzabhängigkeit der Ausbreitung.** Mehrband-Messkampagnen werden mit Richtantennen durchgeführt, die eine omnidirektionale Abtastung der Ausbreitungsumgebung vornehmen. Während der Messungen werden die Tx-Rx-Systeme an festen Positionen platziert und die Ausbreitungsumgebung bleibt so statisch wie möglich. Mit Hilfe von synthetisierten omnidirektionalen Verzögerungs-Leistungsprofilen soll untersucht werden, ob es eine Frequenzabhängigkeit in der Mehrwegeausbreitungsstatistik gibt, z.B. in der Verzögerung und der Winkelspreizung.

(ii) **Einfluss der Antennenrichtwirkung auf die Definition des Kanalmodells.** Es werden Messungen des schnellen Schwunds durchgeführt, die ein Szenario emulieren, bei dem eine Funkverbindung über ein einzelnes Mehrwege-Cluster aufgebaut wird, das mit Antennen mit unterschiedlichen Strahlbreiten ausgeleuchtet wird. Das Hauptziel ist hier die Untersuchung des Einflusses der räumlichen Filterung auf den schnellen Schwund aufgrund der hohen Antennenrichtwirkung. Insbesondere wird die Auswirkung auf Variationen der Empfangssignalstärke und die Gültigkeit der Annahme der schmalbandigen Stationarität im weiteren Sinne (sowohl im Zeit- als auch im Frequenzbereich) untersucht.

(iii) **Einfluss der Systembandbreite auf die Funkkanalmodellierung.** Messungen des schnellen Schwunds werden verwendet, um Mehrwege-Cluster in einem Hörsaal-Szenario auszuleuchten. Das primäre Ziel ist es, den Einfluss einer hohen Systembandbreite auf die Variationen der Empfangssignalstärke, die Zufälligkeit des Kreuzpolarisationsverhältnisses und die Reichhaltigkeit der Mehrwegstreuung zu untersuchen.

Basierend auf der Charakterisierung realistischer Funkkanäle führen die in dieser Dissertation vorgestellten Ergebnisse zu dem Verständnis, dass beim Übergang zu höheren Frequenzen die Frequenz x selbst keine signifikante Rolle bei der Definition der Kanalmodellierungsmethodik spielt. Vielmehr ist es von grundlegender Bedeutung, wie ein Ausbreitungskanal ausgeleuchtet wird. Daher zeigt sich, dass mmWave-Systemeigenschaften wie eine hohe Antennenrichtcharakteristik und Systembandbreite einen hohen Einfluss auf die Definition des Kanalmodells haben. Im Allgemeinen ist die Skalierung der Schwundtiefe als Funktion der Systembandbreite ziemlich gut verstanden. Wir zeigen, dass die hohe Antennenrichtwirkung von mmWave-Systemen zu einer weiteren Reduzierung der Schwundtiefe führt. Zusätzlich erforschen wir einige neue Richtungen in diesem Forschungsbereich, die auf der Analyse der Statistik zweiter Ordnung des Kanalimpulsantwort-Vektors basieren. Unsere Ergebnisse unterstreichen, dass die Schwund-Statistiken der auflösbaren Kanalabgriffe in einem mmWave-Funkkanal nicht als Rayleigh-Rice-verteilte Zufallsvariablen modelliert werden können. Dies liegt vor allem daran, dass durch die hohe Antennenrichtwirkung und Bandbreite von mmWave-Systemen Kanäle mit spärlichen Streubedingungen ausgeleuchtet werden. Folglich ist die Annahme komplexer Gauß'scher Zufallsvariablen, die mit Rayleigh-Rice Schwundverteilungen verbunden ist, nicht mehr gültig. Des Weiteren wird gezeigt, dass die hohe Antennenrichtwirkung und Bandbreite von mmWave-Systemen auch die Gültigkeit der Annahme von Stationarität im weiteren Sinne im Slow-Time-Bereich von mmWave-Funkkanälen in Frage stellt. Die in diesem Beitrag vorgestellten Ergebnisse sind neuartig und bieten theoretisch konsistente Einblicke in den gemessenen Funkkanal.

Contents

Abbreviations

GSM	Global System for Mobile communications
FCC	Federal communications commission
mmWave	milli-meter Wave
CIR	Channel Impulse Response
CTF	Channel transfer function
LTE	Long Term Evolution
NR	New Radio
SOTA	State-Of-The-Art
DSP	Digital Signal Processing
DBF	Digital BeamForming
HBF	Hybrid BeamForming
ABF	Analog BeamForming
RF	Radio Frequency
5G-NR	5G New Radio systems
DAC	Digital-to-Analog-Converter
3GPP	3rd Generation Partnership Project
GBSCM	Geometry Based Spatial Channel Model
MPC	Multi-Path Component
NYU	New York University channel model
MIMO	Multiple-Input and Multiple Output
IO	interacting object or reflection surface
Tx	Transmitter
Rx	Receiver
i.i.d.	independent-and-identically-distributed
XPR	Cross- polarization power ratio

XPD	Cross- polarization power discrimination
SOS	Sum-of-Sinusoids
SOC	Sum-of-Cisoids
AOD	Angle-Of-Departure
AOA	Angle-Of-Arrival
PDP	Power Delay Profile
LOS	Line Of Sight
NLOS	Non- Line Of Sight
UWB	Ultra- Wide Band
PSD	Power Spectral Density
WSS	Wide-Sense Stationary
US	Uncorrelated Scattering
TWDP	Two Wave with Diffuse Power
FTR	Fluctuating Two Ray
SIR	Signal to Interference Ratio
BS	Base Station
MS	Mobile Station
DS	Delay Spread
ASA	Angular Spread of Arrival
ASD	Angular Spread of Departure
LCR	Level-Crossing Rate
AFD	Average Fade Duration
BER	Bit Error Rate
ACF	temporal Auto-Correlation Function
CCF	temporal Cross-Correlation Function
CACF	Complementary Auto-Correlation Function
AWGN	additive white Gaussian noise
RV	Random Variable
PDF	Probability Distribution Function
CDF	Cumulative Distribution Function
KLD	Kullback-Leibler Distance

I	Inphase component of a complex baseband signal
Q	Quadrature component of a complex baseband signal
OFDM	Orthogonal Frequency Division Multiplexing
SCM	Spatial Channel Model
LSP	Large-Scale Channel parameter
V2V	Vehicle to Vehicle communication
HPBW	Half Power Beam Width
QuaDRiGa	Quasi deterministic radio channel generator

Chapter 1

Introduction

In the last 20 years, the number of devices communicating over the internet has grown tremendously as shown by the data obtained from International Telecommunication Union (ITU) in Fig. 1.1. This has led to a massive increase in the data traffic over the radio communication links and a shortage of available spectrum in the sub-6 GHz frequency bands. With limited options left in terms of available spectrum, investigations on propagation characteristics of mmWave *radio channels*[1] have been intensified in the last 10 years. In general, mmWave frequencies are known for their poor link budget properties, which include

- higher attenuation and absorption losses resulting in a higher path loss,
- high diffraction [1], and shadowing [2] losses.

Higher attenuation losses due to small wavelengths can effectively be compensated with highly directional antennas. This can be done by incorporating more antenna elements into the aperture which allows for higher array gains. Due to the small wavelengths, size of antenna elements for mmWave systems are supposed to be small. Thus allowing to build large-scale Multiple-Input

[1] In this thesis, the term *radio channel* always corresponds to the case when an antenna is considered as a part of the channel, whereas a *propagation channel* does not include any antenna effects on the multipath gains.

and Multiple Output (MIMO) systems which provide a high antenna directivity that helps to mitigate higher path losses at mmWave frequencies.

Consequently, both path loss and mmWave channel impulse response (CIR) become antenna dependent. This is different from sub 6-GHz systems, where due to the usage of omni-directional or sectorial antennas, path loss and CIR have very little dependence on the antenna configuration [3]. Thus, mmWave channel models are supposed to have high sensitivity to the antenna directivity.

Millimeter wave systems are also supposed to have large absolute bandwidth which is available at these frequency bands. These features, i.e., large absolute bandwidth and high antenna directivity, enable mmWave systems to be a perfect physical layer technology for future data-hungry wireless networks. However, the success of mmWave systems is strongly coupled with an in-depth understanding of propagation characteristics of mmWave frequency bands and the accuracy of corresponding radio channel models, leading to an accurate performance prediction of a particular mmWave system design.

1.1 Millimeter Wave MIMO Systems

In general, state-of-the-art (SOTA) mmWave MIMO systems can be classified into *analog*, *hybrid* and *fully digital* transmit and/or receive beamforming architectures. For simplicity, discussions in this chapter will focus only on the transmit architectures, unless otherwise stated. The major implementation difference between analog, hybrid, and digital transmit architectures is *where the spatial pre- processing is applied in the overall physical transmitter chain.* In the analog beamforming, the spatial processing is applied only in the radio frequency (RF) domain. Hybrid beamforming architectures require spatial processing in both baseband and in the RF domain, whereas digital beamforming refers to the case where precoding is applied only in the digital baseband unit.

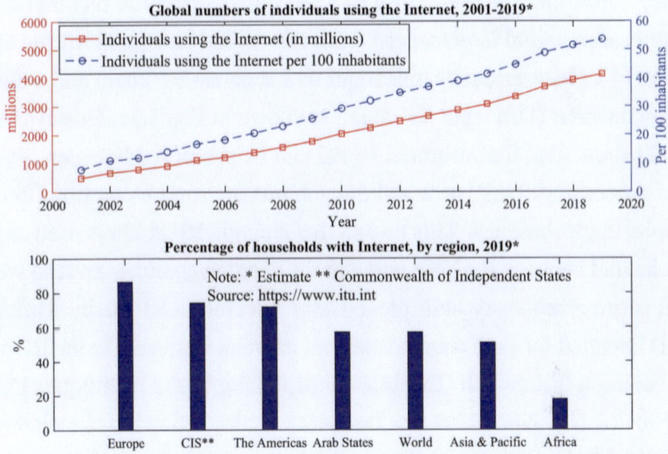

Fig. 1.1. Internet traffic growth and percentage of households with internet connection. Data in these figures is taken from the website of ITU.

Both digital and hybrid beamforming schemes have the potential to transmit multiple data streams.

Digital Beamforming (DBF) Systems

Given the same time-frequency resource (bandwidth), digital beamforming (DBF) enables simultaneous frequency-multiplexed transmission of multiple orthogonal data streams, as shown in Fig. 1.2a. Digital beamforming is optimal in terms of spectral efficiency because it enables transmission of multiple spatially orthogonal data streams (each from an independent beam) per subcarrier.

[2]By definition, the optimality of DBF comes from the fact that digitally pre-processed signal is preserved from the digital baseband till the antenna aperture [4]. Conventionally, this requires a separate RF chain and a digital-to-analog-converter (DAC) per RF chain, as shown in Fig. 1.2a. However, studies in [5–7] show that, the weighted signal can be preserved between the digital signal processing (DSP) unit and the antenna aperture using time/frequency multiplexing techniques. This means that a single RF chain is used as a common channel between the DSP unit and the antenna aperture, and the weighted signal is time/frequency multiplexed over a common RF chain. Multiplexed signals intended for each antenna element are then recovered in the RF domain, either using a fast switch (for time-multiplexing based techniques [5, 6]) or using analog bandpass filters (for frequency-multiplexing based technique [7]). The major implementation challenge for time-multiplexing techniques in [5, 6] is the switching speed requirement for the RF chain to antenna mapper. In practice, this switching speed should be higher than the Nyquist sampling rate, which in this case, is equivalent to the product of bandwidth and number of antenna elements in the transmit array. This poses a serious restriction on the application of DBF techniques in [5, 6] for 5G mmWave MIMO systems due to the requirement on a large number of antenna elements and larger operating bandwidths.

Analog Beamforming (ABF) Systems

Considering the downlink transmission direction, in the analog beamforming (ABF), all frequency resources (bandwidth) are allocated to a single user and multiple users are served in a time-multiplexed way [8]. In contrast to conventional DBF multi-antenna systems, both amplitude and phase weightings of a signal in the ABF are carried out in the RF domain. RF complexity and power consumption in ABF are significantly lower than DBF, because it only

[2]Notice that, time- and frequency- multiplexing concepts discussed in this paragraph are the same as the ones known from wireless communication. However, the application of time- and frequency- multiplexing is restricted only to multiplex over the radio frequency chain, which serves as a common channel between the baseband and the antenna aperture.

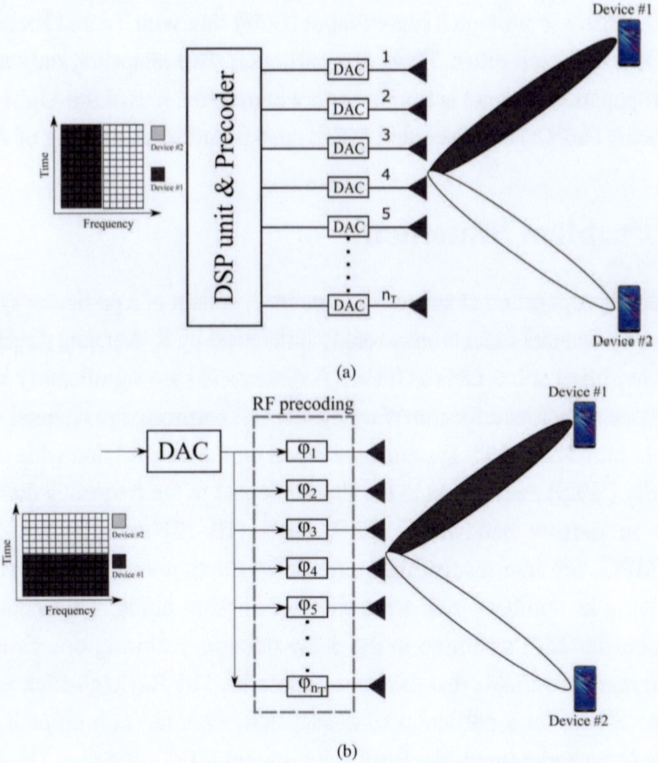

Fig. 1.2. A comparison of potential transmit digital and analog beamforming architectures, (a) Simultaneous (frequency multiplexed) digital beamforming in multiple directions. (b) Non-simultaneous (time multiplexed) analog beamforming in multiple directions.

requires a single RF chain and a single DAC, as shown in Fig. 1.2b. Therefore, in terms of power consumption in the 5G New Radio (5G-NR) systems, the ABF is envisioned to be a better choice for large-scale mmWave MIMO systems, at least in the short- and medium-term perspectives [8].

In ABF, only one data stream can be transmitted through a single beam pointed towards a certain direction in space. A communication link between the transmit (Tx) and receive (Rx) units can be established either through a multipath cluster or through a line-of-sight (LOS) link with Tx and Rx beams pointed towards each other. Thus, at a particular time snapshot, only a part of the propagation channel is illuminated; whereas, the rest of the Multi-Path Components (MPCs) are discarded due to spatial multipath filtering of ABF.

1.2 Problem Statement

While actual propagation characteristics are independent of a particular system design, radio channel models are strongly influenced by it. Antenna directivity and bandwidth of sub 6-GHz 3GPP-LTE systems [9] are significantly lower than the one envisioned for mmWave systems. Therefore, the channel models [10–12] for 3GPP-LTE systems, are based on the narrowband wide-sense stationarity (WSS) assumption in the slow-time and in the frequency domains.

Due to narrow bandwidth, the models [10–12] assume that intra-cluster MPCs are non-resolvable in the delay-domain and they are significantly large in number—rich scattering assumption holds. Consequently, all intra-cluster MPCs sum up in the delay domain, resulting in a complex Gaussian random variable that can be well described by Rayleigh-Rice fading envelops. Since, for a particular time-snapshot, each tap (a multipath cluster) is an independent-and-identically-distributed (i.i.d.) complex Gaussian random variable. Therefore, CIR vector becomes WSS in the slow-time domain, which corresponds to the uncorrelated scattering (US) in the Doppler frequency-domain.

Due to a lower spatial resolution of 3GPP-LTE systems, the number of illuminated MPCs is significantly large. Thus, the central limit theorem holds and each sub-carrier in the frequency domain becomes a complex Gaussian random variable. Multiple sub-carriers are correlated in the frequency-domain up to the coherence bandwidth of a multipath channel. Since the antenna directivity is low, the coherence bandwidth is small due to high temporal-dispersion of MPCs at sub 6 GHz frequency bands. Consequently, the channel

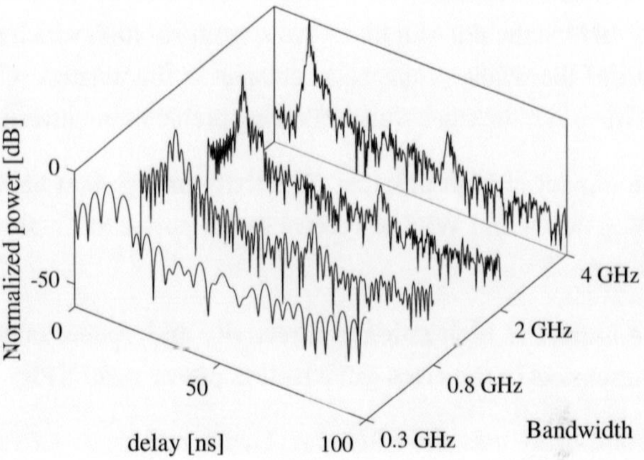

Fig. 1.3. CIR for various absolute bandwidths. Each PDP is normalized locally to its peak power P level i.e. $\sup\{P(\tau)\}$, τ is the delay of an MPC.

models [10–12] are WSS in the frequency-domain, which corresponds to US of resolvable MPCs in the power delay profile (PDP).

In contrast to the 3GPP-LTE systems, the absolute bandwidth available for mmWave systems is supposed to be large, resulting in a high temporal resolution in the delay-domain, as shown in Fig. 1.3. Consequently, the number of non-resolvable MPCs per delay bin are reduced. Due to sparsity in the number of MPCs per delay-resolvable tap, the amplitude envelope of this tap may not be modelled with Rayleigh-Rice fading distributions [13]. Notice that highly directive mmWave beamformers are supposed to illuminate only a part of the whole propagation channel. Therefore, in an extreme case, spatial filtering of a multipath channel may illuminate only a single and potentially sparse multipath cluster or just a single intra-cluster MPC. In such a case, the propagation channel parameters may not be modelled as a random process. Therefore, for an accurate performance prediction of a particular mmWave system design, an in-depth characterization of resolvable channel taps and their modelling becomes a study of fundamental interest. The primary **focus** of this dissertation is exactly the characterization of radio channel when both directivity and bandwidth of a mmWave system are high which means that only a part of the whole propagation channel is illuminated. Considering these mmWave system aspects, the following problems are investigated:

1. **The impact of high antenna directivity and system bandwidth on fading depth and WSS of receive signal power in the slow-time domain.**

2. **The impact of high antenna directivity and system bandwidth on randomness in the cross-polarization power ratio XPR.**

3. **Second-order statistics: Temporal auto- and cross- correlation (between in-phase and quadrature components) properties of the complex CIR in the slow-time domain**.

4. **WSS both in frequency and in the slow-time domain**.

Our **motivation** to study the above mentioned problems comes from the following observations in the SOTA channel models for mmWave systems: Each resolvable multipath component in the CIR vector is an i.i.d. complex Gaussian random variable, e.g.: Each multipath in

(i) 3GPP standardized channel [14], IEEE 802.11ad/ay [15, 16], MiWEBA [17, 3], conference room [18] and METIS [19] models is a complex Gaussian random variable which follows Rayleigh fading distribution,

(ii) New York University (NYU) model [20] is a complex Gaussian that follows Rice fading distribution.

This assumption leads to an i.i.d. complex Gaussian CIR vector, resulting in Rice and Rayleigh multipath channels which are used to describe LOS and NLOS propagation scenarios, respectively. The –high level– contribution of this thesis is as follows:

When the directivity and bandwidth of a mmWave system are high, the CIR cannot be modelled as an i.i.d. complex Gaussian random vector.

1.3 Highlights of Contributions

This thesis -mainly- aims to provide experimental insights into the radio channels illuminated using mmWave systems with high antenna directivity and bandwidth. The contributions of this thesis are:

(i) **Frequency dependency of propagation:** Frequency does not have a significant impact on the delay and angular dispersion statistics of MPCs in an omni-directional propagation channel [21, 22]. This conclusion is drawn under the assumption that absolute bandwidth (same delay resolution), Tx-Rx positions, antenna directivity (physical or synthesized), and the dynamic range of the system remain same for each frequency band. If these assumptions are satisfied, it is quite plausible that the same scatterers are illuminated for two entirely different frequency bands. Due to differences in the wavelength, MPC attenuation losses may differ between the frequency bands. For example, a potentially smooth surface at lower frequency bands may not be smooth at higher frequency bands. However, this may only impact the standard deviation of delay and angular spreads while maintaining the same average value. In this work, we have provided parameter values for a well known empirical model to derive delay and angular dispersion statistics of the omni-directional channels across the frequency bands ranging from 4-73 GHz in different propagation scenarios.

(ii) **Fade depth analysis and modelling:** Fade depth of the receive signal power is an important system parameter as it determines the probability of outage [23]. In this work, an empirical model is proposed that describes fade depth as a function of antenna directivity and bandwidth. Our work [24–26] demonstrates fade depth reduction with an increase in antenna directivity, whereas, it reduces exponentially with an increase in system bandwidth. This is intuitively well expected because the MPCs, towards which, a highly directive antenna beam is pointed to are much stronger in amplitude as compared to other MPCs in the radio channel. In an extreme case, one can envision that only a single MPC is illuminated. Consequently, the impact of weaker MPCs (spatially filtered) on receive signal power fluctuation vanishes with an increase in antenna directivity. Fade depth reduction with bandwidth comes from the fact that multipath delay resolution increases hyperbolically with system bandwidth. Consequently, MPCs in the PDP get delay resolved and each resolvable delay bin becomes temporally sparse leading to reduced power fluctuations in the overall receive signal power.

(iii) **Modelling the phase of resolvable channel taps:** State-of-the-art channel models for mmWave systems [3, 14–20] assume that phase of each resolvable tap is a uniformly distributed random variable over $[0, 2\pi]$ — Rayleigh fading assumption. Theoretically, this assumption cannot be justified when mean value (ρ) of the tap fading process is non-zero as shown in the Fig. 1.4. Additionally, the phase of such a tap converges to be a deterministic process either with an increase in ρ or with a reduction in its standard deviation (σ_0). Notice that the same conclusion holds even when the number of scattered MPCs (N) falling into a resolvable tap are large e.g. in Rice fading channels [27]. A direct consequence to fade depth reduction with an increase in antenna directivity and bandwidth (mentioned earlier) is that the resolvable tap amplitudes may scale more with link-level attenuations and less from the small-scale fading effects. Consequently, the mean of resolvable channels taps in a small-scale fading interval are non-zero and they have small variances. Consequently, the phase distribution of resolvable channel taps look very much differently compared to SOTA channel models as shown in Fig. 1.4. Multipath modelling efforts in this work employ sum-of-sinusoids principle proposed in [27] using ρ and σ_0 values obtained from measurements.

(iv) **Modelling of the polarization coupling matrix:** In narrowband channels, each entry of the polarization coupling matrix is modelled as an i.i.d. Gaussian random process. Same property is maintained in SOTA channel models for mmWave systems e.g. in [14–17, 19, 20]. Deterministic amplitude and phase variations of resolvable taps lead towards the deterministic fading behaviour of co- and cross-polarized channels in the 2×2 polarization coupling matrix. Similar to fade depth, our work [28] shows that randomness in the XPR also reduces exponentially with bandwidth. Therefore, entries of the polarization coupling matrix should be modelled as correlated processes because otherwise channel models would provide an overestimate of the polarization diversity order as shown by the work of Diego et. al. in [29]. Additionally, in SOTA channel models, average XPR remains same for each reflection surface reflecting a cluster of MPCs. We demonstrate that the signal depolarization properties of each surface differ significantly. Our experiments show up-to 10dB differences in the average XPR values which explains how parameters of each individual multipath gain importance from channel modelling perspectives. In this work, we have provided average XPR values and their variances as model parameters for different reflection surfaces.

(v) **Multipath modelling:** State-of-the-art channel models [3, 14–20] for mmWave systems model the resolvable channel taps as complex Gaussian random variables—a narrowband assumption. In narrow-band channels, the number of MPCs falling into each resolvable delay bin is significantly high. Consequently, the complex Gaussian random variable assumption becomes realistic. Our work [30, 31] demonstrates that the complex CIR is a circularly symmetric non-Gaussian random vector and correlation distance is significantly larger than Rayleigh-Rice multipath channels. This is due to the fact that for high bandwidth UWB channels, the number of non-resolvable MPCs (N) becomes significantly small, and resolvable taps do not remain complex Gaussian random variables if $N < 10$ [13]. Same is true also when a propagation channel is illuminated with antenna directivity. In our work [32] we have employed SOS principle introduced in [27] using ρ and σ_0 obtained from measurements. Our results in [32] demonstrate that when a radio channels are illuminated with high antenna directivity then the narrow-band fading statistics of measured channels fit better when $N < 4$ compared to the Rayleigh-Rice channels for which $N \geq 10$.

(vi) **Validity of WSS assumption:** When an Rx moves in a small-scale local area, then, for systems with high antenna directivity and bandwidth, the complex CIR vector does not remain WSS in the slow-time domain [34]. However, in the frequency domain, it can still be modelled as WSS random process. This implies that LOS and NLOS radio channels cannot modelled as stochastic Rice and Rayleigh multipath channels which are inherently WSS in both slow-time and frequency domains. In order for a random process to be WSS, it must be mean and delay invariant. In the absence of fading due to large bandwidth and high antenna directivity, the amplitude of each MPC may scale only with propagation distance. So the mean value (e.g. power/amplitude of the tap) may not be constant even for minuscule movements of Tx and Rx movements. For the complex CIR vector to be delay invariant, it should satisfy the following properties: (a) The auto-correlation functions (ACF) of in-phase and quadrature components should be equal, (b) The cross-correlation function (CCF) between in-phase and quadrature components should either be an odd function or it must be zero-valued. Notice that, variances of in-phase and quadrature components in a sparse multipath channel are not the same [13]. This variance defines value of the ACF between in-phase and quadrature components at a time-lag zero. So the first requirement for delay invariance (property (a)) cannot be satisfied in the

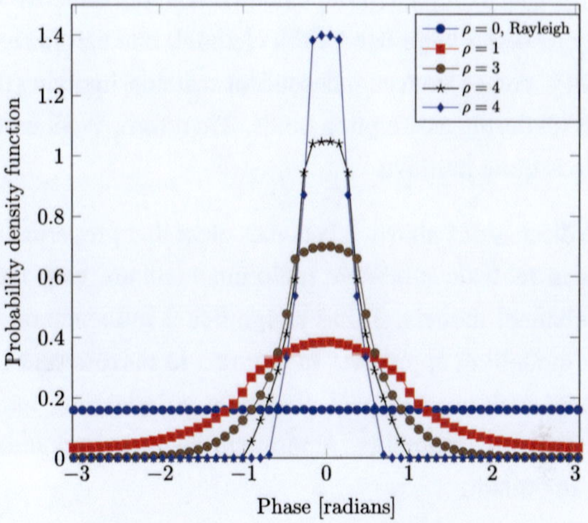

(a). Impact of mean value (ρ) when variance $\sigma_0 = 1$.

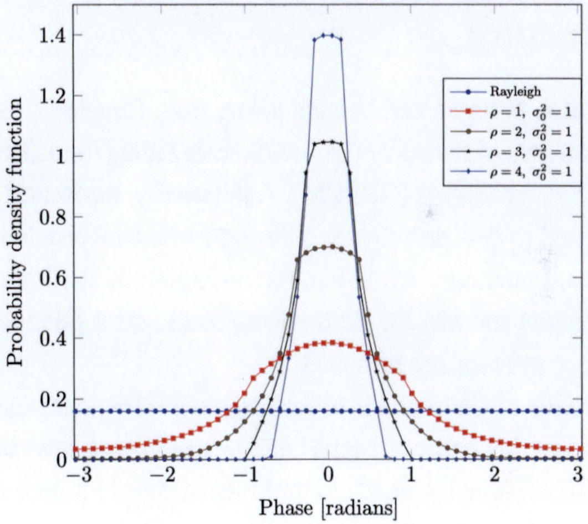

(b). Impact of variance (σ_0) when mean $\rho = 1$.

Fig. 1.4. Phase distribution analysis of a tap when number of MPCs falling into the tap are $N = 3$. Analysis is carried out using well known sum-of-cisoids/sinusoids principle [33]. It is assumed that channel is composed of one strong non-fluctuating MPC having an amplitude ρ and N scattered MPCs having power $2\sigma_0^2$. Further it is also assumed that mean of the phase process is zero.

case of sparse multipath illumination with highly directive antennas. For a delay resolved large bandwidth channel, one can fairly assume that each MPC comes from an independent scatterer implying that the uncorrelated scattering assumption holds. Therefore, WSS in the frequency domain is quite intuitive.

From the discussions above, it becomes clear that properties of resolvable channel taps in realistic mmWave radio channels are very much different than SOTA channel models. It has a significant influence on the accuracy of multipath modelling approach. In contrast to narrowband channel models, parameters such as multipath phase, its polarization, number of non-resolvable MPCs per channel tap along with its mean and variance become significantly important.

1.4 Structure

This dissertation is organized in such a way that, Chapter 2 reviews basics about the main area of the study (i.e. small-scale fading) and Geometry Based Stochastic Channel Models (GBSCM). Additionally, theoretical insights into the models used to describe cluster fading statistics are provided along with their generation methods. After that, the multipath modelling methodology of SOTA channel models for both narrowband sub-6 GHz and mmWave communication systems are reviewed.

The objective of Chapter 3 is to investigate the frequency dependence of propagation. For this purpose, multi-band measurements are carried out at 7, 30, and 60 GHz frequency bands in three scenarios: 1) indoor lecture room, 2) side-link, and 3) typical factory environment. Analysis of multipath delay and angular spreads from these measurements demonstrates a limited impact of frequency on multipath dispersion statistics. Since measurements were carried out over an UWB range, intra-band frequency dependency is also been investigated to analyse the validity of uncorrelated scattering assumption.

In Chapter 4, fading properties of a multipath cluster are investigated using experimental data for single and multi-bounce clusters. Second-order statistical analysis of measured multipath radio channel data demonstrates that when the bandwidth is high then each resolvable tap in the delay-domain cannot be modelled as a complex Gaussian random variable. It is also shown that fade depth and randomness in the XPR vanishes exponentially with an increase in bandwidth.

The objective of Chapter 5 is to show that, mmWave system features such as high antenna directivity and bandwidth play a significant role in defining the channel modelling methodology. It has been shown that, apart from the system bandwidth, the fade depth of a cluster also reduces significantly with an increase in spatial bandwidth. The typical channel modelling assumption of WSS in slow-time, known from narrow-band channels, does not remain valid. It is also shown that the length of stationary interval also depends upon system bandwidth.

Finally, Chapter 6 concludes this dissertation along with a future outlook of this work.

Chapter 2

Fundamentals and Review of State-of-the-art Channel Models

This chapter introduces the main area of study and the necessary background required for a better understanding of the problem which has been addressed in this document. Section 2.1 briefly describes general areas of mmWave channel modelling research and reviews major results from literature compared to microwave frequencies. In Section 2.3, fundamental concepts of multipath modelling approaches have been described for the most commonly used geometry-based spatial channel model (GBSCM) for MIMO simulations. The objective of Section 2.4 is to compare small-scale fading models available in the literature that can potentially be used to describe marginal tap statistics in a GBSCM model. In particular, it is shown that, if non-resolvable MPCs per channel tap are sparse then the fading envelope of the resolvable channel tap is modelled as a circularly symmetric non-Gaussian complex random variable (RV). In Section 2.5, fading waveform generation methods are described for both rich and sparse multipath channel taps. Finally, in Section 2.6, the application of the GBSCM concept in state-of-the-art channel models for MIMO simulations is demonstrated. More specifically, the review of these models highlights multipath modelling approaches in terms of their fading statistics.

2.1 Basic Components of mmWave Channel Modelling

This section briefly describes some of the main areas of research in the mmWave radio channel modelling literature. So far, to the best of our knowledge, mmWave radio channels modelling efforts have focused on the research areas shown in Fig. 2.1. In the following, each one of them is described briefly along with major results found in the literature.

Path Loss (PL).

Assuming that antenna gain remains the same, path loss at mmWave frequencies is high as described by Frii's equation;

$$P_{\mathrm{Rx}} = G_{\mathrm{Tx}} G_{\mathrm{Rx}} P_{\mathrm{Tx}} \left(\frac{\lambda_c}{4\pi d} \right)^2 \tag{2.1}$$

where P_{Tx} is the transmit and P_{Rx} is the receive signal power, G_{Tx} is the transmit and G_{Rx} is the receive antenna gain, λ_c is the wavelength of carrier frequency and d corresponds to the distance between Tx and Rx. Typically, the path loss at mmWave frequencies is around 20 dB higher than microwave channels [35]. However, this can be effectively compensated with an increase in antenna gain of mmWave system.

Shadow Fading (SF).

SF at mmWave frequencies is quite large. For example, results in [2] demonstrate a 30 dB higher shadowing loss caused by human bodies in the 60 GHz band as compared to microwave channels. SF is typically modelled as a log-normal random process.

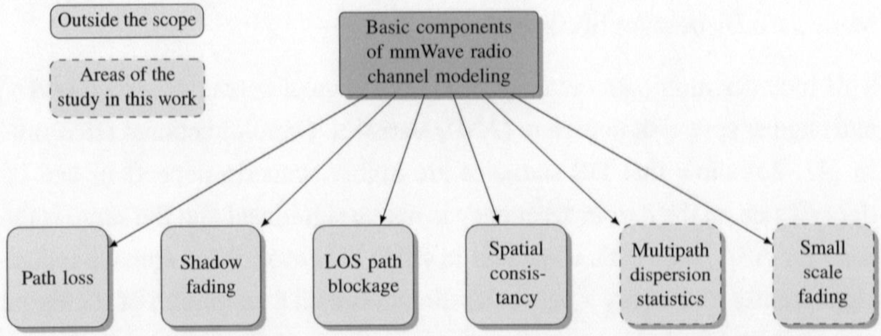

Fig. 2.1. General areas of the mmWave wireless channel modeling research

LOS Path Blockage.

Due to a severe blockage loss, multipaths in a mmWave radio channel may completely/partially be obstructed. It is quite likely that LOS path is blocked/shadowed, even in the case when Base Station (BS) and Mobile Station (MS) have perfect optical visibility with each other. This phenomenon has modified the classical definitions of LOS and NLOS scenarios. In the mmWave channels, LOS and NLOS scenarios are defined based on the probability of LOS path blockage [36].

Spatial Consistency.

Spatial consistency means that spatio-temporal characteristics of a channel smoothly evolve when either Tx, Rx or both of them move in a certain direction [19]. Thus channel snapshots on the neighboring geographical Tx/Rx locations should be correlated. This requires a consistent evolution of both large- and small-scale channel parameters when either of the Tx/Rx (or both) move. A spatially consistent channel model is thus important for an accurate performance prediction of beam tracking algorithms.

Multipath Dispersion Statistics.

This includes multipath delay spread (DS), angular spread of arrival (ASA) and angular spread of departure (ASD) statistics. Omni-directional DS results in [21, 22] show that DS statistics are highly scenario dependent and its dependence on the carrier frequency is not so significant and the same is the case for ASA and ASD, as shown in [21]. However, these spreads reduce significantly when only a particular direction with a maximum of Rx signal energy is illuminated.

Small-Scale Fading.

A minuscule change in the position of either Tx/Rx (or both) or a small variation in the propagation environment changes the delay and phase relationship of multipaths in the PDP. Therefore, MPCs arriving within the spatial and temporal resolution of a system interfere; either constructively or destructively. This results in some random variation of the Rx signal magnitude, leading to small-scale fading and system outage [37]. In contrast to the other studies mention in Fig. 2.1, very few results are available on the small-scale fading behaviour of mmWave radio channels. This contribution aims to fill this gap as much as possible.

2.2 Double-Directional Radio Channels

When there are multiple antennas at both transmitter and at the receiver, then angular distribution of energy at antennas also needs to be modelled. Thus *space* comes as an additional dimension along with *time* and *frequency* modelling of a multipath channel. Such channels are known as double-directional radio channels [38, 39], where

- the term *directional* indicates that model includes angular distribution of multipath energy at the antennas;

- the term *double* correspond to a situation when multipath angles of departure and arrival distributions have been modelled at both transmit and receive antenna arrays.

Assuming that there are N multipath components in the wireless channel, then for a particular position of transmitter and receiver, the narrowband double-directional channel impulse response can be written as

$$h(\tau, \mathbf{\Omega}_t, \mathbf{\Omega}_r) = \sum_{n=1}^{N} h_k(\tau, \mathbf{\Omega}_t, \mathbf{\Omega}_r) \tag{2.2}$$

where τ, $\mathbf{\Omega}_t$ and $\mathbf{\Omega}_r$ respectively represent the delay, the direction-of-departure (DOD) and the direction-of-arrival (DOA) in three dimensional space. Next Section 2.3 provides further theoretical details of double-directional radio channels.

2.3 Geometry Based Spatial Channel Model (GBSCM)

In this section, a generalized description of the GBSCM channel model is provided. Notice that, the GBSCM model forms the basis of many SOTA channel models for MIMO simulations; examples include 3GPP model [10], WINNER [11] and COST2100 [12] models for sub 6 GHz systems and IEEE 802.11ad/ay [15, 40], MIWEBA [3], Conference room [18], NYU [20] and 3GPP-TR 38.901 [14] for systems operating at mmWave frequencies. In general, GBSCM models assume that MPCs arrive and depart in the form of clusters such that their arrival times, departure and arrival angles are similar (see Fig. 2.2). Considering an NLOS scenario, the mth channel tap of the model shown in Fig. 2.2 can be mathematically described as

$$H_{u,s,m}^{\mathrm{NLOS}}(t) = \sqrt{\frac{P_m}{N}} \sum_{n=1}^{N} \boldsymbol{F}_{\mathrm{Rx}}^{T} \underbrace{\begin{bmatrix} \exp(\theta_{m,n,vv}) & \sqrt{\kappa_{m,n}^{-1}} \exp\left(\theta_{m,n,vh}\right) \\ \sqrt{\kappa_{m,n}} \exp\left(\theta_{m,n,hv}\right) & \exp\left(\theta_{m,n,hh}\right) \end{bmatrix}}_{\substack{2\times 2 \text{ Polarization coupling matrix} \\ \text{Phase change due to interaction with } n^{\text{th}} \text{ scatterer}}} \boldsymbol{F}_{\mathrm{Tx}}$$

$$\cdot \exp\left\{ -j\frac{2\pi}{\lambda_0} \left(\underbrace{\hat{r}_{\mathrm{Rx},m,n}^{T} \cdot \bar{d}_{\mathrm{Rx},u}}_{\substack{\text{Phase change due to the} \\ \text{location of Rx antenna element} \\ \text{on the array}}} + \underbrace{\hat{r}_{\mathrm{Tx},m,n}^{T} \cdot \bar{d}_{\mathrm{Tx},s}}_{\substack{\text{Phase change due to the} \\ \text{location of Tx antenna element} \\ \text{on the array}}} + \underbrace{\hat{r}_{\mathrm{Rx},m,n}^{T} \cdot \bar{v}t}_{\substack{\text{Phase change due to the} \\ \text{movement of Rx}}} \right) \right\} \quad (2.3)$$

where,

- N is the number of non-resolvable MPCs falling within a resolvable channel tap,
- P_m is the average power of the resolvable channel tap,
- $\boldsymbol{F}_{\mathrm{Rx}} \in \mathbb{C}^{2\times 1}$ and $\boldsymbol{F}_{\mathrm{Tx}} \in \mathbb{C}^{2\times 1}$ are the vectors corresponding to Tx and Rx antenna field patterns,
- $\theta_{m,n}$ is the phase of the nth ray reflected from m^{th} cluster,
- $\kappa_{m,n}$ is the cross-polarization power ratio (XPR) of the nth ray reflected from m^{th} cluster,
- $\hat{r}_{\mathrm{Tx},m,n}^{T}$ and $\hat{r}_{\mathrm{Rx},m,n}$ are the unit vectors for certain departure and arrival angles (relative to the LOS path) of nth ray reflected from m^{th} cluster,
- $\bar{d}_{\mathrm{Tx},s}$ and $\bar{d}_{\mathrm{Rx},u}$ are the location vectors corresponding to the Tx antenna element s and the Rx antenna element u, respectively.

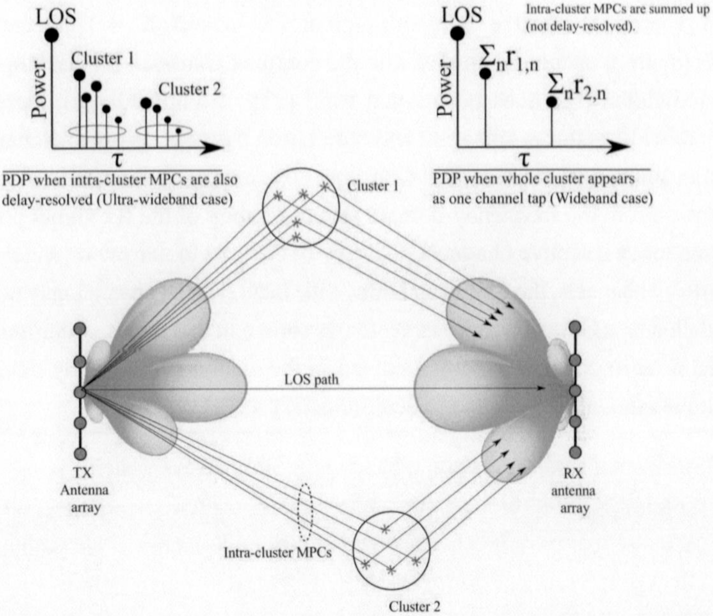

Fig. 2.2. A generalized description of GBSCM model for MIMO simulations, showing a Line-Of-Sight (LOS) path, multipath clusters randomly positioned in space and intra-cluster Multi-Path Components (MPCs).

For wideband systems, the GBSCM modelling methodology considers that multiple intra-cluster MPCs add up and thus result in a single resolvable channel tap in the delay domain, as shown in Fig. 2.2. In such a case, when Rx moves, a change in the delay and phase relationship of MPCs falling within a resolvable channel tap results in an amplitude fading of the tap. If N is significantly large, the resulting tap is modelled as a complex Gaussian RV following a particular fading distribution and the CIR becomes a complex Gaussian random vector.

When the temporal resolution of a system increases with bandwidth, MPCs are resolved in the delay domain. Hence, N reduces for each individual resolvable delay bin, but the number of resolvable channel taps increases. As

N approaches the sparse multipath regime, i.e. towards $N \rightarrow 1$, the central limit theorem cannot be applied and the complex Gaussian RV assumption for individual taps does not remain valid [13]. Additionally, an increase in bandwidth reduces smearing and correlation between resolvable channel taps in the complex CIR [41]. Consequently, increased stochastic degrees of freedom in the frequency-domain reduces fading of the Rx signal power in frequency selective channels [42, 43]. In contrast to narrow or wideband multipath channels, the CIR in an Ultra- wide band (UWB) channel may not be modelled as a Gaussian random vector, as shown in this thesis. Additionally, when a radio channel is fully resolved in the spatial and/or delay domain, multipath modelling becomes a deterministic process.

> Since antenna directivity and bandwidth of mmWave systems is supposed to be high (both spatial and temporal resolution are high), we expect mmWave radio channels with no (or minor) small-scale fading effects.

2.4 Review of Selected Small-scale Fading Models

Small-scale fading is caused by the superposition of both deterministic/non-fluctuating (e.g. LOS and specular reflections) and scattered/diffuse MPCs in the propagation channel. Assuming that there are L non-fluctuating MPCs in a channel, the complex-valued baseband signal can be described as

$$\zeta(t) = \underbrace{\sum_{l=1}^{L} \rho_l e^{j\theta_l}}_{\text{Contribution from non-fluct. MPCs}} + \underbrace{u_1(t) + ju_2(t)}_{\text{Contribution from scattered MPCs}} \tag{2.4}$$

where, ρ_l and θ_l correspond to the amplitude and phase of non-fluctuating MPCs. Additionally, $u_1(t)$ and $u_2(t)$ are RVs whose values are

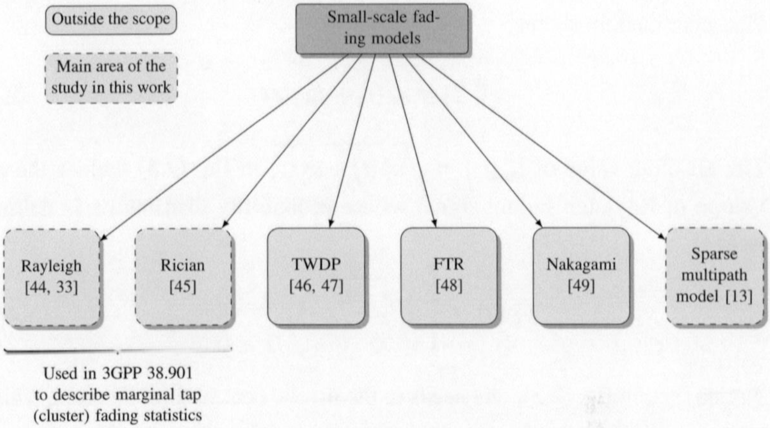

Fig. 2.3. SOTA small-scale fading models

time t dependent. The generalized model in Eq. (2.4) forms the basics of each model presented in this section. Each model shown in Fig. 2.3 demonstrates different properties in terms of the number of non-fluctuating MPCs L, their amplitudes, phase and random distributions of $u_1(t)$ and $u_2(t)$.

The basic objective of this section is to demonstrate that, how properties of a sparse multipath channel differ from conventional models which are based on the rich scattering assumption?

Note that, apart from the *sparse multipath model* in Fig. 2.3, all the models correspond to the rich scattering environment.

2.4.1 Rayleigh and Rice Fading Models

Assuming that there exists no deterministic MPC in the channel i.e., $L = 0$. Additionally, both $u_1(t)$ and $u_2(t)$ are real-valued, WSS, zero-mean mutually uncorrelated Gaussian random processes having equal variance $\left(\sigma_0^2\right)$, then the model in Eq. (2.4) results in a zero-mean circularly symmetric complex-valued

Gaussian random signal

$$\zeta(t) = u_1(t) + ju_2(t)\cdot \tag{2.5}$$

The absolute value of $|\zeta(t)| = \sqrt{u_1^2(t) + u_2^2(t)}$ in Eq. (2.5) defines the envelope of Rayleigh fading signal whose probability distribution is defined as [44]

$$p_\zeta(z) = \begin{cases} \frac{z}{\sigma_0^2} e^{-\frac{z^2}{2\sigma_0^2}} & , \quad z \geq 0\,. \\ 0 & , \quad z < 0 \end{cases} \tag{2.6}$$

Notice that, in Eq. (2.5), one needs to the invoke central limit theorem, which states that $\zeta(t)$ tend to be a complex Gaussian random process with zero mean and variance $2\sigma_0^2$, if the number of scattered multipaths $N \to \infty$. This implies that a Rayleigh fading channel corresponds to rich scattering environment.

If a non-fluctuating multipath component (e.g. LOS path) of amplitude ρ is present in the channel, then Eq. (2.4) simplifies to

$$\zeta(t) = \rho e^{j\theta_\rho} + u_1(t) + ju_2(t)\cdot \tag{2.7}$$

where θ_ρ correspond to the phase of non-fluctuating MPC. Eq. (2.7) results in a non-zero mean Rice distributed random process with a probability distribution function defined as [50]

$$p_\zeta(z) = \begin{cases} \frac{z}{\sigma_0^2} e^{-\frac{z^2+\rho^2}{2\sigma_0^2}} I_0\left(\frac{z\rho}{\sigma_0^2}\right) & , \quad z \geq 0 \\ 0 & , \quad z < 0 \end{cases} \tag{2.8}$$

where $I_0(\bullet)$ is a modified Bessel function of the first kind with zeroth order. The K-factor of a Rician fading channel is defined as the ratio of the power of non-fluctuating path ρ^2 to the power of scattered MPCs $2\sigma_0^2$ [51]

$$K = \frac{\rho^2}{2\sigma_0^2}\cdot \tag{2.9}$$

Since both $u_1(t)$ and $u_2(t)$ are WSS random processes, the auto-correlation function (ACF) of the complex valued random process in Eq. (2.5) is defined as

$$\begin{aligned} c_{\zeta\zeta}(\Delta t) &= E\{\zeta^*(t)\zeta(t+\Delta t)\} \\ &= E\{(u_1(t)-ju_2(t))(u_1(t+\Delta t)+ju_2(t+\Delta t))\} \\ &= \begin{matrix} E\{u_1(t)u_1(t+\Delta t)\}+E\{u_2(t)u_2(t+\Delta t)\} \\ +jE\{u_1(t)u_2(t+\Delta t)\}-jE\{u_2(t)u_1(t+\Delta t)\} \end{matrix} \\ &= c_{II}(\Delta t)+c_{QQ}(\Delta t)+j(\gamma_{IQ}(\Delta t)-\gamma_{QI}(\Delta t)). \end{aligned} \tag{2.10}$$

Since for both Rayleigh and Rice fading channels, $u_1(t)$ and $u_2(t)$ are mutually uncorrelated random processes, implying that $\gamma_{IQ}(\Delta t)=0,\ \forall\Delta t$. Consequently, Eq. (2.10) results in a real-valued temporal function

$$\begin{aligned} c_{\zeta\zeta}(\Delta t) &= c_{II}(\Delta t)+c_{QQ}(\Delta t) \\ &= 2\sigma_0^2 J_0(2\pi f_{\max}\Delta t) \end{aligned} \tag{2.11}$$

where $f_{\max}$ is the maximum Doppler frequency and J_0 is a zeroth-order Bessel function of first kind. The real-valued temporal ACF in Eq. (2.11) explains the reason for a symmetric U-shaped Doppler frequency spectrum of Rayleigh-Rice fading channels.

2.4.2 Two Wave with Diffuse Power (TWDP) Scattering Model

In contrast to the Rician fading channel, the TWDP model assumes two non-fluctuating multipaths ($L=2$) each having amplitude $\rho_1 \geq 0$ and $\rho_2 \geq 0$. However, similar to Rician fading model, the rich scattering assumption is maintained in order to generate two mutually uncorrelated WSS random processes $u_1(t)$ and $u_2(t)$. Therefore, the Rician fading channel can be realized as a special case of the TWDP model. The complex-valued baseband signal for the TWDP fading model is defined as

$$\zeta(t) = \sum_{l=1}^{2} \rho_l e^{j\theta_l} + u_1(t) + ju_2(t) \cdot \tag{2.12}$$

The K-factor for the TWDP fading model is defined as

$$K = \frac{\rho_1^2 + \rho_2^2}{2\sigma_0^2}. \tag{2.13}$$

The model assumes that both non-fluctuating MPCs may have different amplitudes. Therefore, for curve fitting, the model relies on an additional factor Δ, which defines a power relation between both non-fluctuating MPCs in the following way:

$$\Delta = \frac{2\rho_1\rho_2}{\rho_1^2 + \rho_2^2}. \tag{2.14}$$

Note that $0 \leq \Delta \leq 1$ and $\Delta \to 1$ when the amplitudes ρ_1 and ρ_2 are equal.

2.4.3 Fluctuating Two Ray (FTR) Model

FTR model introduced in [48] is a further generalization of the TWDP model. Results in [48] show that, it fits better to the mmWave fading channels as compared to the classical Rice fading model. Recalling Eq. (2.4), similar to the TWDP model, the FTR model assumes two specular components i.e., $L = 2$. However, in contrast to the TWDP model, FTR assumes that specular components are fluctuating and their amplitudes are modelled as Nakagami distributed RVs. This introduces a further degree of randomness in the FTR model as compared to the TWDP model. Similar to TWDP, Rayleigh and Rician fading models, the FTR model also corresponds to a rich scattering channel, as the contribution from diffuse/scattered MPCs is modelled as a complex-valued Gaussian random process.

2.4.4 Nakagami Fading Model

The probability distribution of the Nakagami fading model is defined as

$$p_\zeta(z) = \begin{cases} \frac{2m^m z^{2m-1} e^{-\left(\frac{m}{\Omega}\right) z^2}}{(m-1)!\Omega^m} & \text{if } m \geq \frac{1}{2}, \quad z \geq 0 \\ 0 & \text{if } z < 0 \end{cases} \tag{2.15}$$

where,

$$\Omega = E\{\zeta^2\}$$

and m is the Nakagami shape factor. Similar to models introduced in the earlier Sections 2.4.1 to 2.4.3, Nakagami fading waveforms requires a generation of colored Gaussian RVs $u_i\,(i = 1, 2 \cdots, 2m)$ each having zero-mean and variance σ_0^2. The Nakagami RV ζ in Eq. (2.15) in terms of Gaussian RVs can be described as [33]

$$\zeta = \sqrt{\sum_{i=1}^{2m} u_i^2} \tag{2.16}$$

Summation of mutually independent Gaussian RVs in Eq. (2.16) reveals that Nakagami fading model also corresponds to rich scattering channels.

2.4.5 Sparse Multipath Model

A multipath channel is said to be sparse when it is characterized by a very small number of scattered MPCs, that is on the order of $N \leq 10$ [13]. Consequently, the central limit theorem cannot be invoked which states that the term $u_1(t) + ju_2(t)$ in Eq. (2.4) tend to be a zero-mean circularly symmetric complex Gaussian random process in the limit when $N \to \infty$. In such a case, both $u_1(t)$ and $u_2(t)$ are mutually uncorrelated. The resulting complex random process $u_1(t) + ju_2(t)$ is WSS which can be fully described by the mean and ACF.

In a sparse multipath channel, the central limit theorem cannot be invoked. Therefore, it is quite intuitive to expect that $u_1(t)$ and $u_2(t)$ are mutually correlated random processes. Additionally, the variances of $u_1(t)$ and $u_2(t)$

may not remain the same. In such a case, the complex random process $u_1(t) + ju_2(t)$ in Eq. (2.4) becomes non-Gaussian. A detailed mathematical description of the properties of a sparse channel is provided in next Section 2.5 (see section on *Sum-of-Cisoids (SOC) Principle*).

> **Remark:** In contrast to rich scattering channels, the complex Rx signal in a sparse multipath channel is a circularly symmetric non-Gaussian RV.

2.5 Generation of Fading Waveforms

From Eq. (2.4), it is clear that contribution from scattered MPCs (i.e., random part of the channel) defines a particular fading behavior of a narrowband channel. This section introduces three methods that can be used to generate fading waveforms introduced in Fig. 2.3. Depending upon the requirement, each method can be extended/modified for the simulation of a particular fading channel. The basic objective of each method is to model the random processes $u_1(t)$ and $u_2(t)$ (defined in Eq. (2.4)) satisfying a particular temporal cross-correlation function (CCF). Notice that, *filter method* introduced in the following Section 2.5.1 does not explain a multipath propagation environment. In contrast, *Sum-of-Sinusoids (SOS)* and *Sum-of-Cisoids (SOC)* successfully demonstrate physical plane wave propagation.

2.5.1 Fading Waveforms Generation Methods

Filter Method.

Filter method requires that an additive white Gaussian noise (AWGN) signal $v_i(t) \sim \mathcal{N}(0,1)$ is applied as an input to a linear time-invariant filter $G_i(f)$, $i = 1,2$, as shown in Fig. 2.4. Filter coefficients are optimized such that, signals at the output of each filter satisfy the properties required by

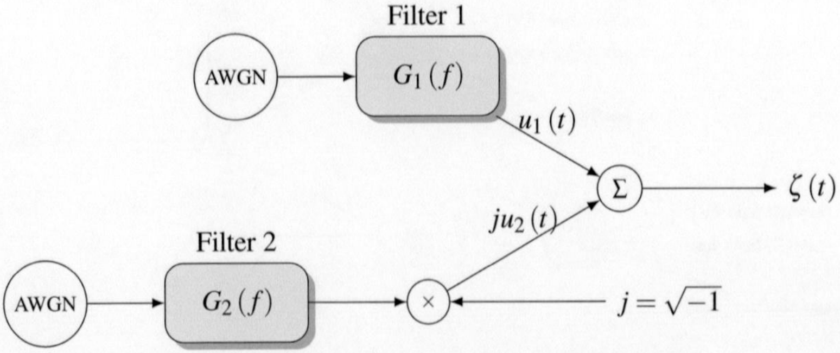

Fig. 2.4. Fading waveform generation using filter method

the RVs $u_1(t)$ and $u_2(t)$ in Eq. (2.4). For example, in the case of rich scattering channels, both $u_1(t)$ and $u_2(t)$ are supposed to be modelled as WSS and mutually uncorrelated colored[1] Gaussian random processes.

Sum-of-Sinusoids (SOS) Principle.

The SOS principle, as the name implies, requires a summation of sinusoids (plane waves) to model the random processes $u_1(t)$ and $u_2(t)$. For the simulation of a particular fading channel, the SOS principle can be mathematically described as [50, 33]

$$u_i(t) = \sum_{n=1}^{N_i} c_{i,n} \cos(2\pi f_{i,n} t + \theta_{i,n}) \tag{2.17}$$

where, N_i in (2.17) is a finite number and it corresponds to the total number of sinusoids, $c_{i,n}$ is the gain, $f_{i,n}$ is the Doppler frequency, and $\theta_{i,n}$ correspond to the phase of a sinusoid. In general, it is assumed that gains $c_{i,n}$ and $f_{i,n}$ $\forall i = 1,2$ are fixed and the $\theta_{i,n}$ is a uniformly distributed random process $\mathcal{U}(0, 2\pi]$. Usually, SOS-based simulators are derived under the assumption that both

[1]Colored Gaussian random process $u_i(t)$ can be considered here as a result of filtering the white Gaussian noise.

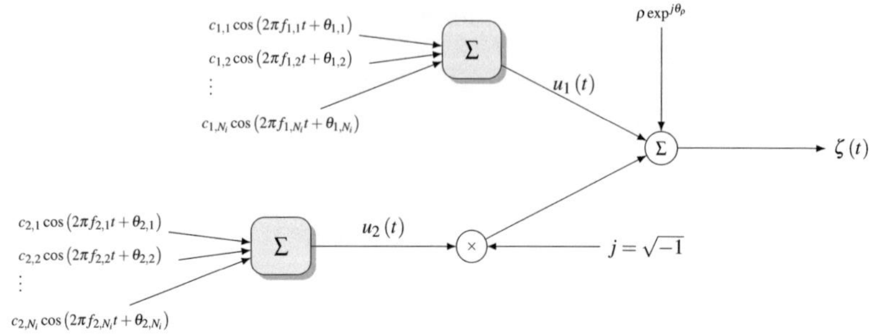

Fig. 2.5. Fading waveform generation using SOS principle

$u_1(t)$ and $u_2(t)$ are mutually uncorrelated random processes—as required by the models in Sections 2.4.1 to 2.4.4. This can be ensured when

$$\{|f_{1,n}|\}\bigcap\{|f_{2,n}|\}=\emptyset \tag{2.18}$$

where, $\emptyset$ corresponds to an empty set. Since both $u_1(t)$ and $u_2(t)$ are mutually uncorrelated, the resulting Doppler PSD would be a symmetric function. This limits the application of SOS-based simulators to multipath channels having an isotropic scattering condition, which is hard to justify, both in the realistic propagation channels [33] and in the radio channels illuminated with directional antennas [45, 52]. Since the CIR in mmWave radio channels is supposed to be antenna dependent (see Section 1.2), the application of the classical SOS principle for the mmWave fading channel simulations is not realistic. Therefore, in the next section, a modified SOS-based approach called SOC principle is presented which can effectively be used to simulate radio channels having asymmetric Doppler PSDs. Assuming that, there exists a non-fluctuating multipath with an amplitude ρ and phase θ_ρ, then Fig. 2.5 provides a graphical description of the SOS-based fading channel simulator.

Sum-of-Cisoids (SOC) Principle.

A special case of SOS-based fading channel simulator, known as SOC principle is obtained by replacing $c_n = c_{1,n} = c_{2,n}$, $f_n = f_{1,n} = f_{2,n}$ and $\theta_n = \theta_{1,n} = \theta_{2,n} + \frac{\pi}{2}$ for all $n = 1, 2, \cdots, N$ in Eq. (2.17), where $N = N_1 = N_2$ corresponds to number of scatterers or diffuse MPCs. An SOC-based model for fading channel simulation is defined as

$$\tilde{u}(t) = u_1(t) + ju_2(t) = \sum_{n=1}^{N} c_n e^{j(2\pi f_n t + \theta_n)}. \tag{2.19}$$

Let us assume that, there exists a single non-fluctuating MPC with gain ρ in Eq. (2.4) and $u_1(t)$ is correlated with $u_2(t)$. Additionally, we consider that MPC gains c_n and Doppler frequencies f_n are constants but their phase θ_n is a uniformly distributed random process $\mathscr{U}(0, 2\pi]$, then PDF of the resulting random process $\zeta(t)$ (in Eq. (2.4)) is defined as [33, see Eq. (4.120)]

$$\begin{aligned} p_\zeta(z) = (2\pi)^2 z \int_0^\infty &\left[\prod_{n=1}^{N} J_0(2\pi |c_n| x)\right] \\ &. J_0(2\pi z x) J_0(2\pi \rho x) x dx, \quad N \geq 1 \end{aligned} \tag{2.20}$$

where J_0 is a zeroth-order Bessel function of first kind. Notice that, the PDF in (2.20) can be used to describe fading statistics of a multipath channel tap when the number of non-resolvable MPCs arriving within the resolution limit of the system are quite low. Let Δt be the time-lag, then the temporal ACF of the random variable $\zeta(t)$ is defined as

$$\begin{aligned} c_\zeta(\Delta t) &= E\{\zeta^*(t)\zeta(t+\Delta t)\} \\ &= \sum_{n=1}^{N} c_n^2 e^{j2\pi f_n \Delta t} \end{aligned} \tag{2.21}$$

From Eq. (2.21), the ACF of inphase $c_{u_1 u_1}(\Delta t)$ and the quadrature components

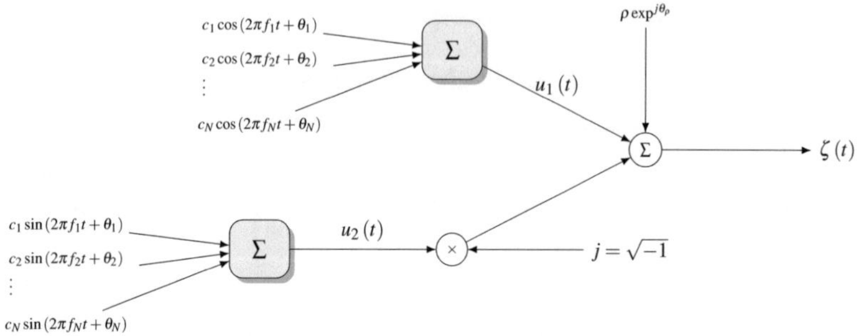

Fig. 2.6. Fading waveform generation using SOC principle

$c_{u_2u_2}(\Delta t)$ of $\zeta(t)$ can be obtained as

$$\begin{aligned} c_{u_1u_1}(\Delta t) = c_{u_2u_2}(\Delta t) &= \frac{1}{2}\mathbb{R}\{c_\zeta(\Delta t)\}, \qquad \mathbb{R} \rightarrow \text{real part} \\ &= \sum_{n=1}^{N} c_n^2 \cos(j2\pi f_n \Delta t) \end{aligned} \tag{2.22}$$

and the CCF function $\gamma_{u_1u_2}(\Delta t)$ of random processes $u_1(t)$ and $u_2(t)$

$$\begin{aligned} \gamma_{u_1u_2}(\Delta t) &= \frac{1}{2}\mathbb{I}\{c_\zeta(\Delta t)\}, \qquad \mathbb{I} \rightarrow \text{Imaginary part} \\ &= \sum_{n=1}^{N} c_n^2 \sin(j2\pi f_n \Delta t)\cdot \end{aligned} \tag{2.23}$$

In contrast to the models introduced in Sections 2.4.1 to 2.4.4, the Eq. (2.23) clearly indicates that I and Q components of the random variable $\zeta(t)$ are correlated for $\Delta t > 0$. At $\Delta t = 0$, the CCF $\gamma_{IQ}(\Delta t) = 0$ in Eq. (2.23) which demonstrates that I and Q components are instantaneously mutually uncorrelated, as shown in Fig. 2.7. However, for $\Delta t > 0$, the random variable $\zeta(t)$ does not satisfy the complex Gaussian RV assumption. Consequently, the ACF of a sparse channel differs considerably from that of Rayleigh–Rice or any other rich scattering channels, as shown in Fig. 2.7a. Similarly, the Rx signal

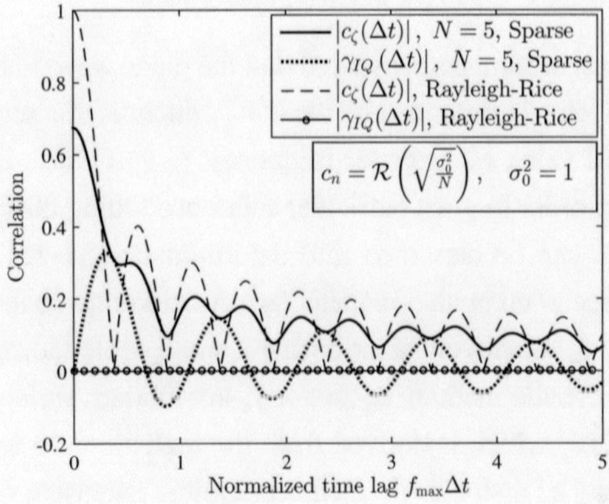

(a)

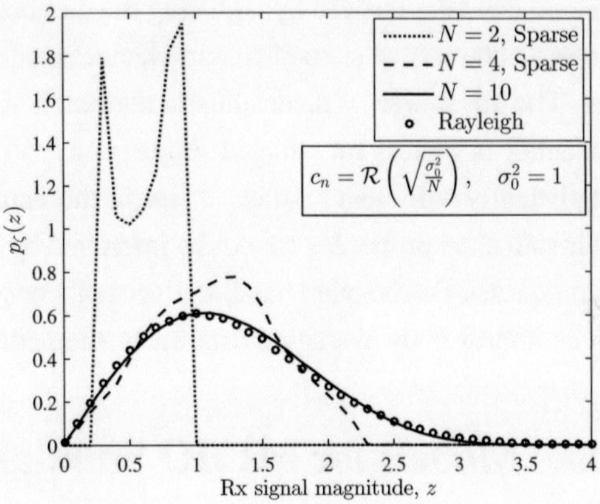

(b)

Fig. 2.7. (a) Analysis of temporal correlation functions and (b) Comp. of PDF in Eq. (2.20) for different L with Rayleigh fading PDF in Eq. (2.6).

magnitude statistics of a sparse channel differ from the Rayleigh–Rice fading channels as shown in Fig. 2.7b.

2.5.2 Parameter Computation Methods.

While the previous section demonstrated that the plane wave multipath propagation can be effectively described by the SOC principle; this section explains, how the sinusoid gains c_n, Doppler frequency f_n and phase θ_n need to be parameterized in order to get a particular reference fading distribution. This parameterization can be classified into deterministic [53–55] and stochastic [56, 57] parameter computation methods[2]. A deterministic method implies that both c_n and f_n remain constant during a small-scale fading interval. In contrast, in a stochastic method, c_n and/or f_n are random variables, obtained from a certain PDF which is derived from the realistic field measurements. Thus the accuracy of a stochastic method requires extensive field measurements. On the other hand, given a certain reference model, the accuracy of a deterministic method can be improved by adjusting the number of sinusoids.

Table 2.1 shows a summary of state-of-the-art channel models corresponding to each class. The advantage of deterministic method is that it provides ergodic fading channel emulators for which a single (or a few) channel snapshots provide statistically sufficient results. It lies in the nature of ergodic channels that their statistical properties cannot be improved by increasing the number of simulation runs. On the other hand, statistics of a non-ergodic simulation model can be improved by averaging over many simulation results [60].

2.6 Channel Models for MIMO Simulations

State-of-the-art channel models for MIMO systems can be classified into stochastic, quasi-deterministic, deterministic and analytical channel models, as shown in Fig. 2.8. Analytical models, characterize the CIR between individual

[2]Note that, in the channel modelling literature, the term *deterministic* is used to describe a variety of concepts. Here, it is used only to describe a method, used for fading waveform generation. This should not be confused with the term *deterministic channels* used later in this dissertation. In this document, the term *deterministic channels* refers to the case when some (or all) non-fluctuating MPCs can be deterministically computed, either from the geometry of an environment (e.g. Ray-tracing) or through any numerical evaluations.

Table 2.1. Classes of intra-cluster MPC modeling methodology and channel models corresponding to each class.

Parameter computation method	Model
Deterministic method (Constant c_n, constant Doppler frequencies[a] f_n and random phases θ_n) Properties: Mean and auto-correlation ergodic [33]	3GPP-TR 38.901 [14]
	mmMAGIC [58]
	METIS-2020 [19]
	WINNER II [11]
	3GPP SCM [10]
Stochastic method (Random c_n, Random Doppler frequencies[b] f_n and random phases θ_n) Properties: Mean ergodic, non-ergodic auto-correlation [33]	IEEE 802.11 ad/ay [15, 40]
	Conf. room model [18]
	MiWEBA [3]
	Extended Saleh-Valenzuela model [59]
	NYU [20]

[a]The models [14, 10, 11, 19, 58] do not explicitly model the Doppler spectrum. However, fixed intra-cluster MPC AOA assignments from the cluster power angular spectrum demonstrate constant Doppler frequencies of sinusoids per snapshot. Note that mmMAGIC and METIS channel models have focused largely on the fast fading parametrization for the 3GPP standardization activities while maintaining the intra-cluster MPC modeling methodology.

[b]Models [3, 15, 40, 18, 59] consider that intra-cluster AOAs are randomly distributed within a cluster angular spread. NYU [20] model assumes MPC clustering in the delay domain only, and MPC may arrive only from certain directions in space known as spatial lobes. Random intra-cluster MPC AOAs per snapshot demonstrate random doppler frequencies of the sinusoids.

antenna elements of the Tx-Rx arrays in terms of mathematical functions without taking into account the electromagnetic wave propagation behavior in a particular environment. Studies in this thesis are based on the statistical characterization of realistic wave propagation in radio channels. Therefore, analytical channel models have been intentionally ignored in this review. For interested readers, a comprehensive survey of analytical channel models is available in [61].

Apart from the analytical models, each model mentioned in Fig. 2.8 falls into the category of GBSCM channel models introduced in Section 2.3. How-

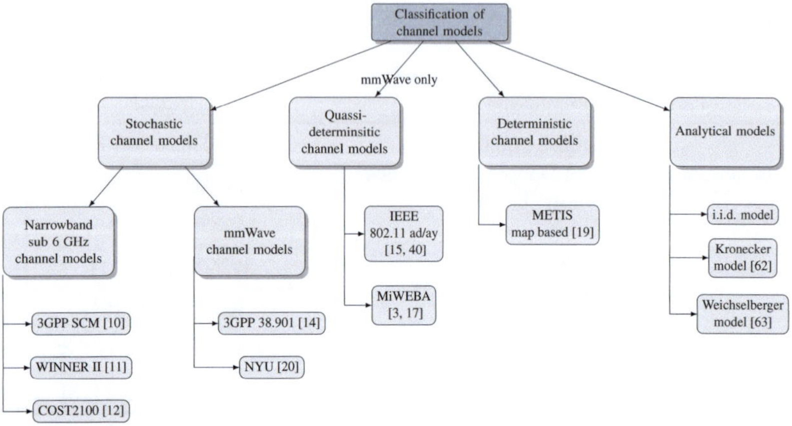

Fig. 2.8. Classification of channel models for MIMO simulations

ever, multipath modelling approach in each model may differ due to the marginal channel tap fading statistics. Generally, a multipath propagation channel comprises of both large-scale (Path loss, diffraction, shadowing) and small-scale (multipath delay, AOD, and AOA) channel parameters. However, to maintain the focus of study, only marginal tap fading statistics and accordingly the multipath modelling approaches of the channel models shown in Fig. 2.8 are reviewed. It is shown that all models employ Rayleigh–Rice fading distributions to describe the marginal tap fading statistics. This implies that taps in the complex Gaussian CIR vector are Gaussian random variables that follow Rayleigh fading distribution when the mean is zero; otherwise Rician.

2.6.1 Channel Models for Narrowband Sub 6 GHz Systems

This section reviews channel models designed specifically for 3GPP-LTE systems [9] which employ small MIMO configurations. Since the antenna directivity is low, path-loss and the complex CIR have a little dependence on the antenna gain [3]. This becomes the primary reason for the rich scattering

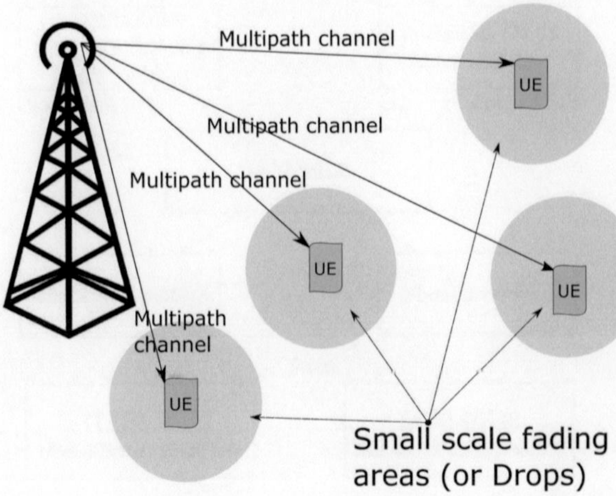

Fig. 2.9. Illustration of drop concept used in multipath channel models. UE corresponds to User Equipment

assumption in [10–12, 64]. These models are based on the drop concept as shown in Fig. 2.9. A drop corresponds to a local area where small-scale fading statistics are applicable. Due to a small bandwidth, the models [10–12, 64] are based on the narrowband WSS assumption. This means that within a drop duration, the complex CIR is WSS in slow-time and in the frequency domain.

3GPP-SCM, WINNER and COST2100 Models.

3GPP-SCM and WINNER: These models belong to the family of double-directional GBSCM channel models and are parameterized for 2–6 GHz carrier frequencies. A summary of evolution from COST 231 (Hata) and COST 231 (Walfish-Ikegami) pathloss models [65] to WINNER II channel model is shown in Figs. 2.10a and 2.10b. Both *link* and *system* level models have been proposed in 3GPP-SCM and WINNER channel models. Link level simulation model is used for calibration purpose to compare the implementation of the same algorithm developed by different research groups/companies. System-level model is used for system wide channel

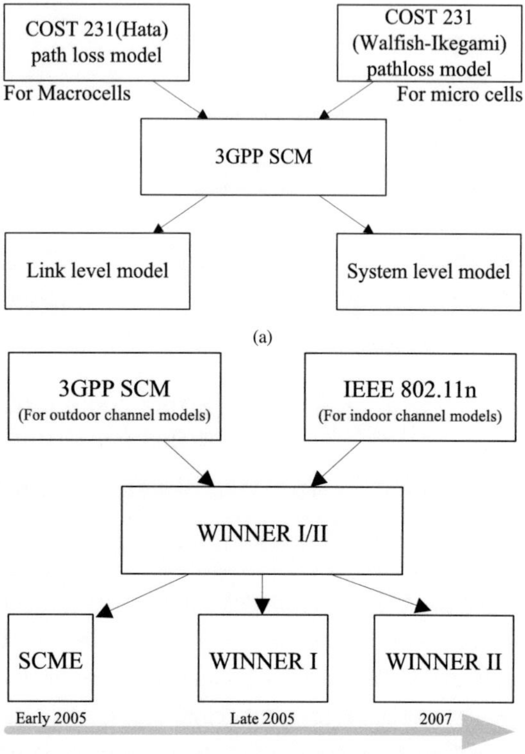

(b)

Fig. 2.10. Evolution of (a) 3GPP-SCM and (b) WINNER channel models (WIM$\rightarrow$ WINNER Interim Model).

simulations. Temporal multipath dispersion in SCM/WINNER channels is defined by 3 types of clusters i.e., Local Scattering Cluster (LSC), Reflecting scatter clusters (RSC), and Far Cluster Scatterer (FSC) which are positioned randomly in space according to a certain PDF [66]. Depending upon a particular scenario, the number of clusters in a channel may vary between 6–20. Since SCM/WINNER model is designed for narrow-band 3GPP-LTE systems, intra-cluster MPCs are not resolvable in the delay domain. Further, it is assumed that intra-cluster number of MPCs are significant enough such that a delay resolvable channel tap can be modelled as a complex Gaussian random variable following Rayleigh fading distribution described in Eq. (2.6). Intra-cluster MPCs are modelled using a deterministic parameter computation method as shown in Table 2.1. This means that MPC gains and AOAs are constants values whereas each channel snapshot is generated by random variation in the MPC phase.

COST-2100 model: COST-2100 [12] channel model assumes local (around the MS), single– and multi– bounce clusters each having a particular visibility region (VR) for the MS. The basic objective of VR (indexed between 0–1) is to ensure a consistent time evolution of the channel for the accurate evaluation of beam-tracking algorithms. In general, COST-2100 model provides a frame work that needs to be parameterized and validated for a particular frequency band using realistic MIMO channel measurements such as in [67, 68]. The amplitude fading envelope of each cluster is a Rayleigh distributed random variable [12]. This is justified by the measurement in [68] which show that the average number of intra-cluster MPCs is considerably high, i.e. around 27 to 48.

Remark: In 3GPP-SCM, WINNER and COST2100 models, modelling of resolvable channel taps as complex Gaussian random variables is quite intuitive because these models are derived for narrowband systems having low antenna directivity. Thus the number of MPCs falling into each resolvable tap are large enough to satisfy the requirements for complex Gaussian random variable.

2.6.2 Channel Models for mmWave Systems

3GPP–38.901 Model.

This 3GPP standardized GBSCM model [14] is an extension of 3GPP-SCM [10], WINNER [11] and COST-2100 [12] MIMO channel models. It aims to provide a consistent evolution of propagation properties from sub 6 GHz to 100 GHz mmWave band. Similar to its predecessors [10–12], 3GPP–38.901 report provides an antenna independent omni-directional channel model and is based on narrow-band assumptions such as WSS in slow-time and in the frequency domain. Channel taps are Rayleigh distributed complex Gaussian random variables and the CIR is a complex Gaussian random vector. Notice that 3GPP has also provided a modelling framework for channels when system bandwidth $B > \frac{c}{D}$ where D [m] is the size of antenna aperture either in azimuth or in elevation direction and c is the speed of light. This framework assumes that when the condition $B > \frac{c}{D}$ is satisfied, a multipath channel can be modelled as a fully delay resolved channel.

Remark: How does the fading properties of resolvable taps evolve in channels which are not fully delay resolved remains an open question in 3GPP–38.901 model. This is primarily due to the fact that the objective of 3GPP–38.901 report is primarily to provide a common platform for different companies to compare their algorithms and it does not fully reflect realistic radio channel properties [69].

QuaDRiGa Channel Model.

Quasi deterministic radio channel generator (QuaDRiGa) model [70] is also an extension of WINNER channel model and it was used as a baseline model for modelling mmWave radio channels in mmMAGIC project [58]. Basic objective of QuaDRiGa modelling methodology is that all large- and small scale channel model parameters evolve deterministically according to the trajectory of user equipment. Modelling of resolvable taps in QuaDRiGa model is same to its legacy WINNER [11] channel model in the sense that these taps are modelled as complex Gaussian random variables following Rayleigh fading distribution. This is primarily due to the fact that summation of 20 non-resolvable sub-paths each having a random phase result in a complex Gaussian random variable—a narrowband assumption. In contrast, findings of this dissertation are different such that we investigate cluster fading properties as a function of antenna directivity and system bandwidth which lead towards sparse multipath illumination. Consequently, resolvable channel taps cannot be modelled as complex Gaussian random variables.

New York University (NYU) Model.

The NYU model in [20] is based on the outdoor measurements carried out in New York city at 28 GHz and 73 GHz frequency bands. Similar to the classical Saleh-Valenzuela model [71], the NYU model assumes that MPCs arrive at Rx in the form of clusters in the delay-domain, as shown in Fig. 2.11a. These clusters are denoted as *time clusters*. However, intra- time cluster MPCs arrive randomly from a few distinct directions in space, as shown in Fig. 2.11b. These directions in space are denoted as *spatial lobes*. A time cluster is modelled in the same way as in the Saleh-Valenzuela model [71] with a maximum of 30 delay-resolved intra-cluster taps each sampled at a minimum 2.5 ns time interval. Each tap is modelled as a complex Gaussian RV and its amplitude fading follows the Rician density function. The origin of this result is the analysis of data obtained from small-scale fading measurement campaign carried out in [72]. Transmit and receive antennas used in this measurement campaign are directional (30° HPBW) and the channel sounder transmits an 800 MHz null-to-null bandwidth signal. Power measurements are carried out in perfect LOS and obstructed LOS scenarios to analyse the power fading of each resolvable tap.

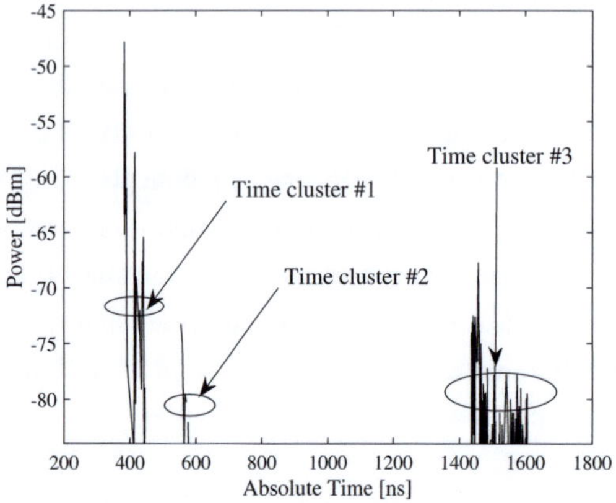

(a)

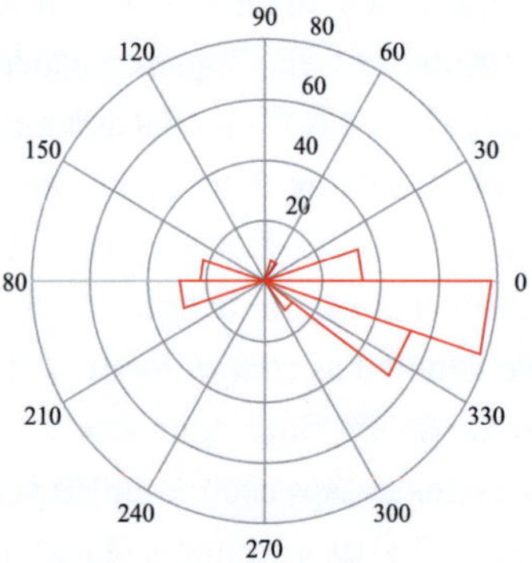

(b)

Fig. 2.11. (a) Power delay profile and (b) MPCs arrival angle histogram of NYU city measurement data.

Remark: In the above-mentioned measurement scenario, modelling of resolvable taps as Rice distributed random variables is questionable because the complex Gaussian RV assumption associated with Rice fading taps is hard to justify. This is mainly due to the fact that one intuitively expects a sparse radio channel illumination with directive antennas and large bandwidth, particularly in LOS scenario.

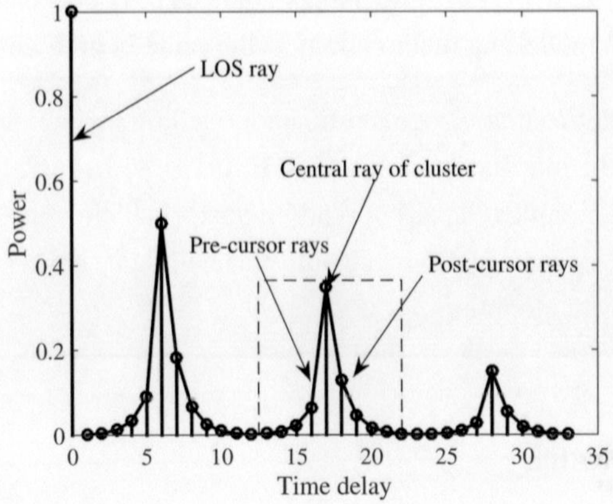

Fig. 2.12. Cluster power delay structure in IEEE 802.11ad channel model

IEEE 802.11ad/ay

IEEE standardized channel models IEEE 802.11ad for indoor scenarios have been proposed for 60 GHz mmWave frequency band with a bandwidth up to 5 GHz [15, 73]. The CIR is composed of both deterministic and stochastic MPCs. Deterministic MPCs such as LOS and some specular reflections from walls are always available in the channel and are derived from ray-tracing simulations. These deterministic MPCs form the central ray of the cluster as shown in Fig. 2.12. Stochastic MPCs (Pre- and post- cursor rays of the cluster) are resolvable in the delay domain and form a cluster structure similar to that of classical Saleh-Valenzuela model [71]. Amplitude decay factor of pre- and post- cursor rays of a cluster may differ from each other: it is generally extracted from channel measurements in a particular scenario. The model assumes that pre- and post- cursor taps in a cluster are complex Gaussian random variables that follow Rayleigh fading distributions.

IEEE 802.11 ay model covers both indoor and outdoor scenarios for the 60 GHz frequency band (5 GHz bandwidth). In contrast to the IEEE 802.11ad model, a cluster in IEEE 802.11ay model contains only post cursor rays: rest of the multipath modelling methodology is the same in both models.

Remark: Notice that, above multipath modelling methodology leads to the conclusion that complex CIR in IEEE 802.11ad/ay channel models is a complex Gaussian random vector which is questionable particularly for such a large bandwidth channel. Same holds for cluster-level multipath modelling.

MiWEBA Model.

MiWEBA (Millimetre-Wave Evolution for Backhaul and Access) is a publicly supported research project to investigate the potentials of 60 GHz frequency with a 250 MHz bandwidth in outdoor access, front- and back-haul communication scenarios. Similar to IEEE 802.11ad channel model, 3D-MiWEBA model [3, 74] is also quasi-deterministic: it contains both deterministic and stochastic MPCs. Deterministic MPCs such as LOS and ground reflections, denoted as D-rays, are always available in the channel. Stochastic MPCs denoted as random (R-) and flashing (F-) rays appear randomly in a channel due to mobility of the MS and motion of humans/cars in the surrounding environment. An example PDP of a MiWEBA model is shown in Fig. 2.13.

Multipath modelling approach in MiWEBA model is same as that of classical Saleh-Valenzuela channel model [71] with multiple clusters in delay domain. Intra-cluster MPCs are zero-mean complex Gaussian RVs$\rightarrow$ Rayleigh fading distribution. Amplitude of intra-cluster MPCs decays exponentially with time as shown in Fig. 2.13. Azimuth and elevation angles of arrival and departure of intra-cluster MPCs follow uniform random distribution within a particular angular range.

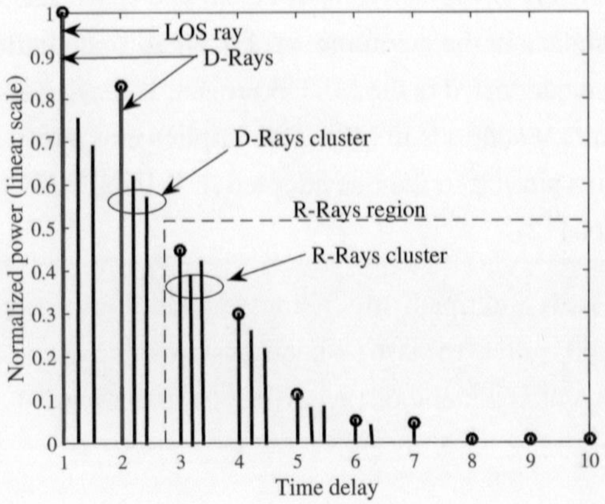

Fig. 2.13. Cluster power delay structure in MiWEBA channel model

METIS Model.

METIS (Mobile and wireless communications Enablers for the Twenty-twenty Information Society) is a European 7th framework project which concluded its channel modelling activities targeted for 5G application scenarios in [19, 75, 76]. METIS identified that standardized channel models [10–12, 15] are inadequate for 5G systems such as massive MIMO, device-to-deice (D2D), vehicle-to-vehicle (V2V) and for mmWave frequency bands. METIS proposed a further extension of WINNER [11] by providing values of large scale channel parameters for the bands between 6-80 GHz carrier frequency. Additionally, map based and hybrid channel modelling approaches are proposed which combine both GBSCM and map-based modelling.

Map-based model is based on a simplified ray-tracing approach using site-specific maps. It is valid for carrier frequencies up to 100 GHz with a relative bandwidth up-to 10% of the center frequency. Unlike the IEEE 802.11ad [15] model METIS channel models consider diffraction modelling an important aspect at mm-waves. Therefore, map-based model characterize large scale parameters as the sum of losses from free space and diffraction loss from multiple objects in the environment. For small scale fading, no explicit studies have been conducted in the METIS project. Instead, it is recommended to use ITU-recommendations in [76]. This implies that multipath modelling approach remains similar to the one adopted in WINNER/SCM[10, 11] and COST200 models[12].

> **Remark:** Thus multipath modelling methodology in the METIS model is well suited to narrow-band systems and does not reflect the properties of channel models having large bandwidth.

2.7 Conclusion

Studies in this chapter demonstrate that the channel taps in state-of-the-art channel models for mmWave systems have been modelled as a complex Gaussian RV following Rayleigh–Rice amplitude fading distributions. This assumption does not remain valid when MPCs falling into a resolvable channel tap are small. This would particularly be the case when a highly directive beamformer illuminates a small part (or only a single MPC) of a multipath channel. It has also been shown that MPCs falling into a resolvable channel tap also become small with an increase in system bandwidth. Consequently, one may expect that high antenna directivity and bandwidth of mmWave systems not only eliminate the reasons for resolvable MPCs to fade but also lead to an understanding that resolvable MPCs should be modelled as sparse taps which are non- complex Gaussian RVs. In such a case, the complex CIR vector shows significantly different properties than the existing state-of-the-art channel models for mmWave systems. This dissertation intends to explore such properties. In the upcoming chapters, it is shown that high frequency itself does not play a significant role in defining the mmWave channel modelling methodology. Instead, mmWave system properties such as high antenna directivity and bandwidth introduce a major difference between the channel models proposed for 3GPP-LTE and mmWave systems.

Chapter 3

Frequency Dependence of Propagation

3.1 Introduction

Conventionally, mmWave bands span from 30–300 GHz frequency range. How does propagation properties evolve when moving up to the higher frequency bands is a question of continued interest; irrespective of channel models at these frequencies which depend upon the system assumptions. This means that, even if propagation channel characteristics are the same for two different frequency bands, channel models may still differ considerably in particular when system properties are different. With an objective to compare the impact of frequency on the propagation channel characteristics, this chapter provides an extensive analysis of large-scale channel parameters using three measurement campaigns in widely different scenarios. The motivation of such an analysis comes from the literature review of temporal dispersion statistics shown in Table 3.1. Notice that, observations made in different measurement campaigns are not consistent in Table 3.1. For this, one may blame differences in the sounding systems, antenna characteristics, scenario, and measurement time for different frequency bands. Therefore, in order to avoid these complications, a multi-band sounder introduced in [77] is developed at the electronic measurements research laboratory at Technische Universität Ilmenau, Germany. Channel sounder is equipped with scanning Tx- Rx horn antennas while maintaining approximately the same antenna directivity (30° HPBW) for different frequency bands. Given the same Tx and Rx positions, the channel sounder can do multi-band simultaneous measurements with the same absolute 7 GHz channel bandwidth. This allows us to fairly compare propagation characteristics of different frequency bands by fulfilling following requirements mentioned in [22]:

- Equal measurement bandwidth (same multipath resolution in delay domain),
- Comparable antenna pattern (either physical or synthesized),
- Same environment ,
- Same antenna locations,
- Static environment.

3.2 Measurement Campaigns

For the channel sounding, two time-synchronized dual-polarized UWB channel sounding units are used which have been introduced in [77]. Sounding signal is a chip sequence of length 4095 with a chip rate of 6.75 GHz. It offers an instantaneous absolute bandwidth of approximately 7 GHz providing a high resolution in the delay domain. Dynamic range is up to 70 dB enabling the identification of very weak MPCs.

3.2.1 Lecture Room Scenario

This is a dual-band measurement campaign conducted at the mmWave band (30.4 to 37.1 GHz) and the FCC-UWB band (3.4 to 10.1 GHz) in a small

Table 3.1. Literature review of DS results for frequency dependency analysis

Scenario	Frequencies	Antennas	Observation
Indoor [78]	1.7 and 60 GHz	1.7 GHz →Omni-directional 60 GHz →20 dBi horn	Directional transmission at 60 GHz results in reduced DS than 1.7 GHz system
Indoor [79]	2 and 60 GHz	2 GHz →Omni-directional dipole 60 GHz →Biconical antenna	50% reduced DS at 60 GHz
Indoor [80]	2.9 and 29 GHz	2.9 GHz →Omni-directional Tx-Rx 29 GHz →Omni-directional	Higher DS at 2.9 GHz in NLOS Higher DS at 29 GHz in LOS
Indoor [81]	28 and 73 GHz	28 GHz → 28.8° HPBW 73 GHz → 15° HPBW	Directional DS is higher in 28 GHz bands
Outdoor [82]	28 and 73 GHz	28 GHz → 10.9° HPBW 73 GHz → 7° HPBW	Directional DS is higher in 28 GHz bands
Outdoor [83]	38 and 60 GHz	38 GHz → 7° HPBW 73 GHz → 7.3° HPBW	Directional DS is higher in 38 GHz bands
Outdoor [84]	10 and 60 GHz	10 GHz → Omni-directional 60 GHz → Omni-directional	Omni-directional DS reduces with an increases in freq.
Indoor [85]	17 and 60 GHz	Omni-directional	No relationship between DS and freq.
Indoor [86]	1 to 11 GHz	Omni-directional	No relationship between DS and freq.
Indoor [87]	850 MHz, 1.9 GHz, 4 GHz, 5.8 GHz	Omni-directional	No relationship between DS and freq.
Indoor [88]	60 and 70 GHz	Synthesized omni-directional	No relationship between DS and freq.
Indoor, outdoor and outdoor-to-indoor (mmMAGIC) [22, 58]	Various bands in the frequency range 6–100 GHz	Synthesized omni-directional	No relationship between DS and freq.

lecture room shown in Fig. 3.1a at the Technische Universität Ilmenau campus. One of the inner- brick walls is covered by metallic blackboards and an opposing wall has 5 cm thick sound absorbers. The outer wall is largely covered by windows with metallic frames. Doors and windows are closed during the measurements, otherwise indicated. Distance to a building opposite the seminar room is approximately 55m. A perpendicular wing of the building next to the seminar room is approximately 19m away. The height of both Tx and Rx units is 1.68m which is approximately an eyesight height. Tx is placed at a fixed location near the blackboard as shown in Fig. 3.1a. However, the Rx unit is moved to 30 different locations inside the room. For both frequency bands, dual-polarized horn antennas with identical 3 dB half-power beamwidth of 30° with 11-12 dBi gains have been used. These antennas are mounted at both Tx and Rx units in completely opposite directions, and they scan 180° opposite directions at the same time for both mmWave and FCC UWB bands. Both spherical positioning devices do a 360° azimuth scan of the lecture room in 12 steps at a particular position. In this way, for 30 different Rx locations inside the room, a total of $12 \times 12 \times 6 \times 5 = 4320$ directional PDPs have been measured. Measurements are conducted in the LOS environment, however, during the azimuth scans, Tx/Rx horn antennas may not be pointed towards each other, hence also illuminating a pseudo-LOS (pLOS) environment. Three PC's are used to control the positioning devices, switches for the polarization, and record the data sent from UWB sounding units.

3.2.2 Vehicle to Vehicle (V2V) Communication Scenario

These tri-band (6.75 GHz, 30 GHz, and 60 GHz) measurements were performed at the "T" intersection in an urban environment with parked cars, multi-story buildings, and lampposts as shown in Fig. 3.1b. The Tx and Rx imitate two cars that are 44 m apart communicating to each other with antennas at a rooftop level with the heights shown in Fig. 3.1b. Two parked cars (labeled Car 1 and Car 2) were present during measurements to increase the scattering effects. A third vehicle, labeled blocker, was positioned at 15 different locations (Position 1-15 shown in Fig. 3.1b), emulating an overtaking situation. This results in LOS path blocking at Positions 5 to 12. In addition to the 15 positions with the blocker vehicle, a reference scenario – without the blocking vehicle in the environment – was also measured. To analyze two typically occurring blocking scenarios, we performed measurements with two blocking vehicles: i) small blocker, represented by a passenger vehicle (Volkswagen Sharan), which was slightly taller than the Tx and Rx antennas of the sounding apparatus; and ii) large blocker, represented by a Mercedes Benz Sprinter van, which was approximately 80 cm taller than the Tx and Rx antennas. Fig. 3.1b shows details of the measurement setup, including dimensions of vehicles. All the measurements were performed in as static as possible scenario by restricting access to streets.

(a). 360° panoramic view of the lecture room

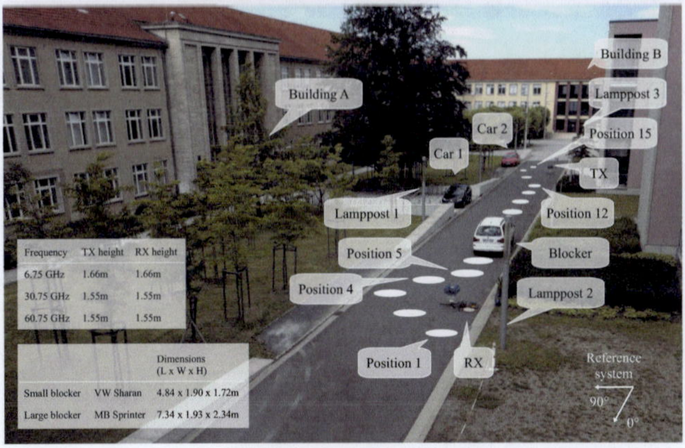

(b). V2V communication scenario

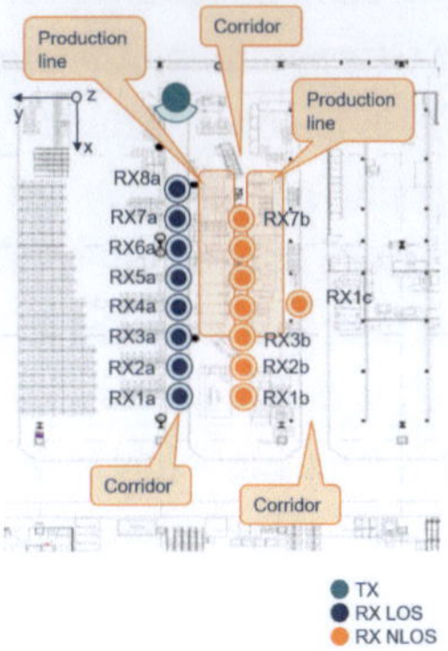

(c). Industry scenario

Fig. 3.1. Overview of measurement scenarios

3.2.3 Industry Scenario

In this measurement campaign, tri-band (6.75 GHz, 30 GHz, and 60 GHz) measurements are carried out in an industry hall at the Bosch factory as shown in Fig. 3.1c. All three UWB measurements are carried out simultaneously using 30° HPBW horn antennas with a multi-band channel sounding setup introduced in [77]. At the Tx (emulating an access point) side, a 180° azimuth and 90° elevation scan is carried out with a step distance of 30°. Similarly, at the Rx (emulating a user equipment) side, a full 360° angular azimuth scan and 90° elevation scan is carried out. Height of Tx is 4m above the ground, whereby, Rx is of height 1.96m. During the measurements, Tx is placed at a fixed position and Rx is positioned at 16 different locations, as shown in Fig. 3.1c. In the case of LOS measurements, a perfect optical visibility between Tx and Rx is maintained. However, in NLOS setups, LOS path to the Rx is obstructed by the production lines, as shown in Fig. 3.1c.

Data Post-Processing

At first the influence of the measurement system is excluded from the recorded channel data sets using the same system calibration procedure as described in [89]. Due to the spectrum band limitations, reduced signal magnitude appears at both edges of the 7 GHz spectrum. Therefore, the whole measurement bandwidth is not usable. These frequency samples are removed from the band by applying time domain gating by using a raised cosine window with a roll of factor of 0.5. It ended up with a 3 dB bandwidth of approximately 4 GHz. At the second stage, a noise floor is determined as 25 dB below the maximum power path and samples below this noise floor are zeroed out. Note that this noise floor value of 25 dB is kept consistent in each directional power delay profile regardless of any directional pointing angle and polarization setup.

3.3 Frequency Dependence in Omni-Directional Channels

In this section, root mean square (rms) delay and angular spreads are compared for measured channels corresponding to different frequency bands. Statistics of rms DS are obtained from synthesized omni-directional propagation channels which provide nearly same results as that of measurements conducted with physical dipole antennas [80]. Let $P(\tau_i)$ be the total power of all MPC arriving at a delay time τ_i, then the first central moment i.e., mean excess delay (MED) $\bar{\tau}$ is defined as,

$$\bar{\tau} = \frac{\sum_i \tau_i P(\tau_i)}{\sum_i P(\tau_i)} \tag{3.1}$$

The second central moment i.e., rms DS is then defined as

$$\tau_{\mathrm{rms}} = \sqrt{\frac{\sum_i (\tau_i - \bar{\tau})^2 P(\tau_i)}{\sum_i P(\tau_i)}}. \tag{3.2}$$

Similarly, assuming that θ_i is an antenna pointing angle towards the ith direction in space, then the angular spread is defined as

$$\sigma_{\mathrm{A}} = \sqrt{\frac{\sum_i (\theta_i - \bar{\theta})^2 P(\theta_i)}{\sum_i P(\theta_i)}} \tag{3.3}$$

where, $\bar{\theta} = \frac{\sum_i \theta_i P(\theta_i)}{\sum_i P(\theta_i)}$ is the average angular spread. Notice that, multipaths are not direction resolved due to 30° HPBW antenna. This requires array measurements and high resolution parameter estimation (HRPE). Therefore, $P(\theta_i)$ in Eq. (3.3) is the power summation of all MPCs along the delay-domain in a particular Tx- Rx directional scan. Consequently, angular spread estimation provided by Eq. (3.3) in our particular measurement setup provides a coarse angular spread estimate. In order to analyze frequency dependence of spatio-temporal dispersion statistics, we use linear regression model in [14, 22] for

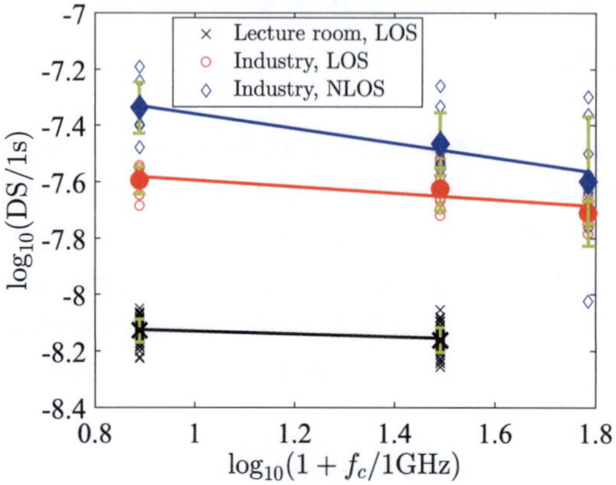

(a). Indoor hotspot scenario

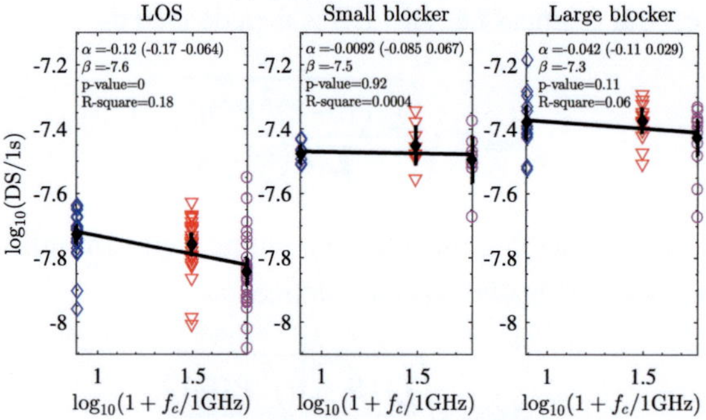

(b). Outdoor V2V communications scenario

Fig. 3.2. Analysis and modelling of temporal dispersion statistics of synthesized omni-directional channels as a function of frequency

the average value:

$$\mu_{\log \mathrm{DS}/1\mathrm{s},\mathrm{ASA}/1^\circ,\mathrm{ASD}/1^\circ} = \alpha \log_{10}(1+f_c)+\beta \tag{3.4}$$

where, α and β are slope and intercept of the linear regression line, respectively. Clearly measurement data samples are not large enough. This is due to limitations of the channel sounding hardware which employs directional

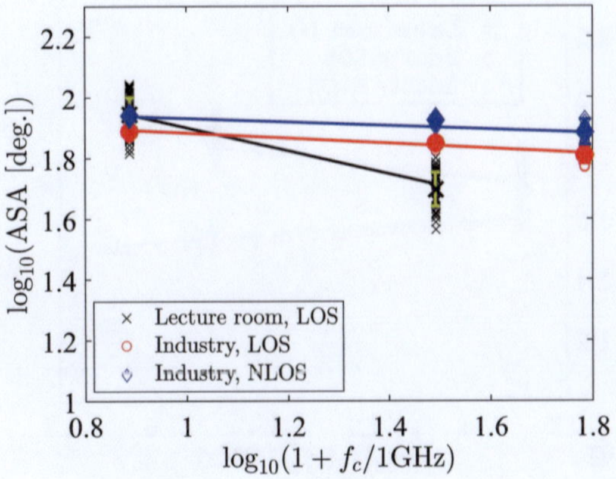

(a). Indoor hotspot scenario

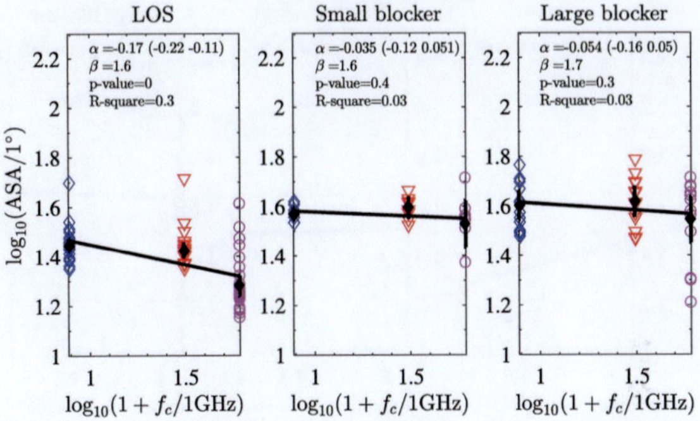

(b). Outdoor V2V communications scenario

Fig. 3.3. Analysis and modelling of angular-spread-of-arrival statistics of synthesized omni-directional channels as a function of frequency

horn antennas to scan the whole environment which is quite tedious and time consuming process. This small number of data samples limit us to draw any statistically significant conclusions. Therefore, in our analysis, p-value and R-square tests are conducted to analyze frequency dependent trend in measured channel parameters. We assume a null hypothesis $\alpha = 0$ for the model in Eq. (3.4) meaning that there exists no statistically significant frequency

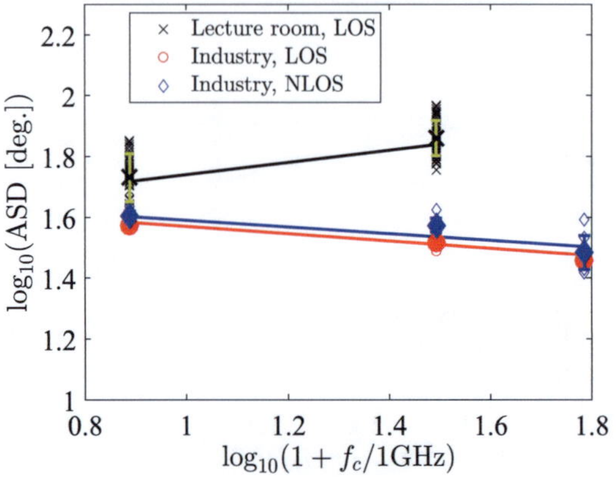

(a). Indoor hotspot scenario

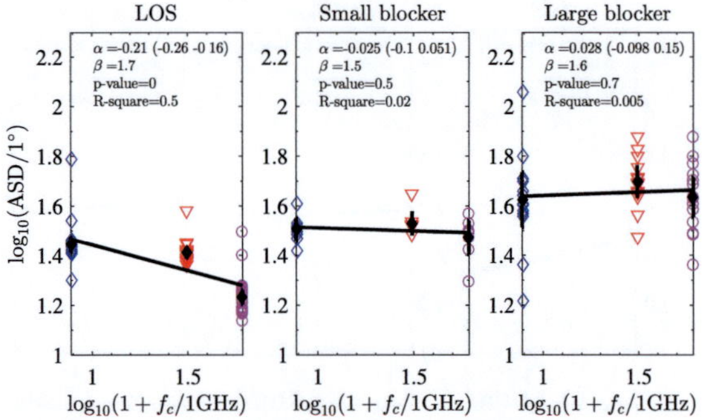

(b). Outdoor V2V communications scenario

Fig. 3.4. Analysis and modelling of angular-spread-of-departure statistics of synthesized omnidirectional channels as a function of frequency

dependency. We assume a p-value threshold i.e., $p \leq 0.05$ to demonstrate that the frequency trend is statistically significant. Additionally, an R-square test is conducted to analyze goodness of the linear regression fit. In our analysis, a linear regression fits pass the goodness of fit test if R-square value is greater than 0.8. Analysis of delay and angular dispersion results in Section 3.3 and Fig. 3.4 provide following interesting insights of the measured channels:

Table 3.2. Parameter values of model in Eq. (3.4) and results of p-value and R-square tests

Scenario	Parameter	α	β	p-Value	R-square
Lecture room LOS	DS	-0.05	-8.08	0 (A)	0.14
	ASA	-0.38	2.28	0 (A)	0.82
	ASD	0.2	1.54	0 (A)	0.47
Industry LOS	DS	-0.12	-7.48	0.004 (A)	0.34
	ASA	-0.08	1.96	0 (A)	0.72
	ASD	-0.12	1.68	0 (A)	0.9
Industry NLOS	DS	-0.26	-7.1	0.007 (A)	0.3
	ASA	-0.05	1.99	0.002 (A)	0.35
	ASD	-0.11	1.7	0 (A)	0.5
V2V LOS	DS	-0.12	-7.6	0 (A)	0.18
	ASA	-0.17	1.6	0 (A)	0.3
	ASD	-0.21	1.7	0 (A)	0.5
V2V with small blocker	DS	-0.0092	-7.5	0.92 (R)	0.0004
	ASA	-0.035	1.6	0.4 (R)	0.03
	ASD	-0.025	1.5	0.5 (R)	0.02
V2V with large blocker	DS	-0.042	-7.3	0.11 (R)	0.06
	ASA	-0.054	1.7	0.3 (R)	0.03
	ASD	0.028	1.6	0.7 (R)	0.005

- Both delay and angular spreads in factory environment are much higher as compared to the typical indoor and outdoor LOS and NLOS scenarios. This is due to a comparatively increased number of reflections in the factory environment.

- Obstruction of LOS path due to vehicle blockage results in an increased spatio-temporal multipath dispersion.

- Size of blocker plays an important role in defining the multipath dispersion statistics—increased dispersion in case of large blockers—demonstrating the need of blocker size as a model parameter.

- In Table 3.2, negative slopes for both delay and angular spreads explain that delay and angular spreads are positively correlated, which agrees with the intuition. Similar results have been shown earlier in [90, 91].

- Frequency dependency of the large scale parameters seems to be statistically significant as confirmed by p-Values in Table 3.2—Null hypothesis is accepted in most of the cases except in the LOS obstruction scenario. However, one may note that, R-square values are significantly lower than 0.8 demonstrating that the linear regression model Eq. (3.4) does not pass goodness-of-fit test due to large standard deviation in measured data samples. Results in Section 3.3 and Fig. 3.4 also show that delay and angular spreads at different frequency bands are quite comparable and there exist no specific trend with frequency. These observations are inline with the mmMAGIC findings shown in [22, 58].

3.4 Frequency Dependence in Sub-bands of Directional Radio Channels

Theoretically, wideband channels could be frequency dependent/selective due to a large difference in wavelengths from lower-to-upper edge frequencies of the band. This means that, each subcarrier will attenuate MPCs differently.

In other words, each subcarrier will see the propagation channel differently. Consequently, fading of each resolvable MPC in a multipath channel become correlated. Therefore, uncorrelated scattering assumption does not remain valid. Studies in [92] show that modeling frequency dependence improves the channel modeling accuracy up-to 40%.

Considering the same absolute bandwidth, in mmWave bands, a difference in the wavelength from lower-to-upper edge frequencies of the band are considerably smaller than microwave frequency bands—relative bandwidth is low. Intuitively, one may expect that, amplitude attenuation of MPCs in mmWave frequency bands will approximately be same across each subcarrier of the whole spectrum. Additionally, since the directivity of mmWave systems is supposed to be high, one may expect lower delay spread and flat (roughly) frequency spectrum. Primary objective of this section is to analyze frequency dependence of mmWave radio channels when antenna directivity is high. For this purpose, dual-band measurements conducted in a small lecture room (see Section 3.2.1) are used. Whole frequency spectrum of directional radio channel corresponding to each Tx-Rx antenna pointing angle is divided into subbands, each with an 800 MHz bandwidth. After that, rms DS statistics are extracted and compared for each subband.

Let $H_d(f)$ be the frequency response of a directional UWB radio channel. For subband channel analysis, $H_d(f)$ is divided into multiple subbands $H_s(f) \in H_d(f)$ each with an absolute bandwidth $\Delta f = f_h - f_l$, where f_h and f_l are the highest and lowest frequencies of the band, respectively. This is equivalent to applying a rectangular filter in the frequency domain. In order to maintain the original length of $H_d(f)$, each subchannel vector $H_s(f)$ is zero padded as

$$H_s(f) = \begin{cases} H_d(f), & f_l \leq f \leq f_h \\ 0 & \text{otherwise.} \end{cases} \tag{3.5}$$

In this way, length of $H_d(f)$ is maintained in each subchannel vector $H_s(f)$. Finally, $h_s(\tau)$ is obtained by inverse Fourier transform of $H_s(f)$ and is used to study multipath delay dispersion statistics of different subbands.

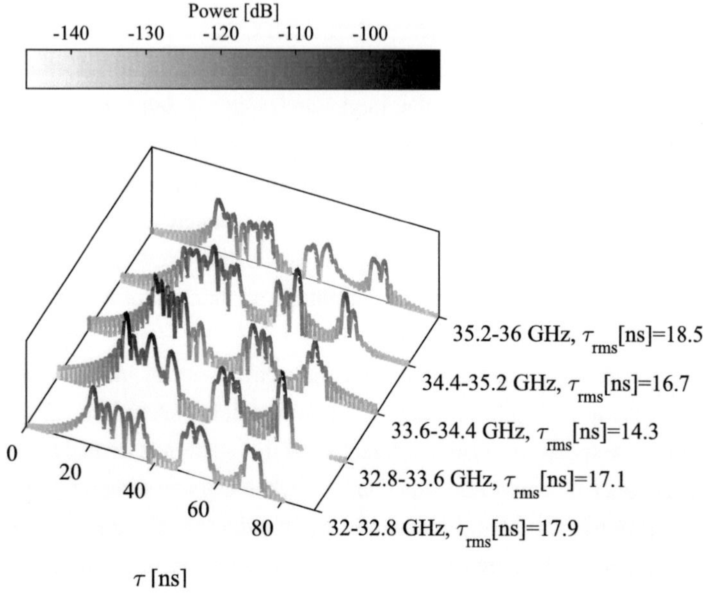

Fig. 3.5. An example of frequency dependent attenuation of MPCs in the PDP

Fig. 3.5 shows example PDPs of measured subband radio channels. Clearly, MPCs are attenuated differently for each subband and result in different rms DS values. Fig. 3.6 shows CDF plots of directional rms DS statistics for each of the measured Tx-Rx polarization setup. These statistics are derived from 4320 measured channel PDPs. Notice that, rms DS in cross polarized Tx-Rx antenna setups is higher than the co- polarized case. More importantly, there exists a certain difference in the rms DS statistics of different subband channels. This shows that rms DS characteristics are not monotonic along the considered mmWave UWB channel and each subband see a different channel due to difference in the wavelength. Similar results can be seen in Fig. 3.7, which correspond to 7 GHz frequency band channels in the same lecture room environment. One may also note that, DS values of directional radio channels

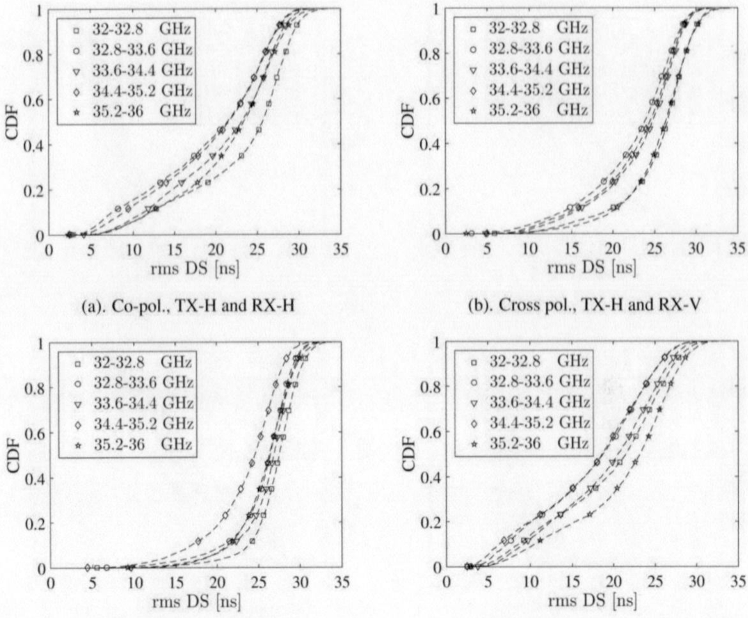

(a). Co-pol., TX-H and RX-H

(b). Cross pol., TX-H and RX-V

(c). Cross pol., TX-V and RX-H

(d). Co-pol, TX-V and RX-V

Fig. 3.6. Directional rms delay spread statistics of different subbands, 34 GHz measurements in a lecture room scenario.

shown in Figs. 3.6 and 3.7 are significantly larger than the omni-directional channels DS values shown earlier in Fig. 3.2a, which seem counter intuitive. Reason behind this observation is that, the bandwidth of each sub-band is five times lower than the full channel bandwidth which causes an increased multipath dispersion in the delay domain. Consequently, rms DS in directional radio channels becomes higher.

Now the objective is to analyze the significance of difference between rms DS statistics of each sub-band. For this purpose, Kullback-Leibler Distance (KLD) is used to quantify the difference between PDFs of each sub-band channel shown in Figs. 3.6 and 3.7. Let P and Q be probability mass functions of rms DS values, then relative entropy $D_{KL}(P \parallel Q)$ a.k.a. KLD quantifies a

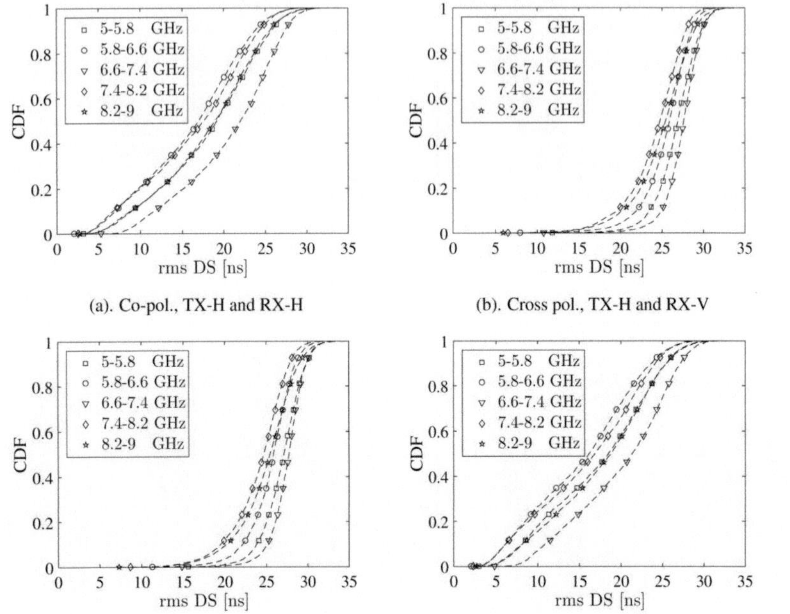

(a). Co-pol., TX-H and RX-H

(b). Cross pol., TX-H and RX-V

(c). Cross pol., TX-V and RX-H

(d). Co-pol, TX-V and RX-V

Fig. 3.7. Directional rms delay spread statistics of different subbands, 7 GHz measurements in a lecture room scenario.

distance between P and Q [93]

$$D_{KL}(P \parallel Q) = \sum_{r} P(r) \log\left(\frac{P(r)}{Q(r)}\right), \tag{3.6}$$

where, r corresponds to a particular rms DS value. Since $D_{KL}(P \parallel Q)$ may not be equal to $D_{KL}(Q \parallel P)$, a symmetric metric is built by taking the average of two opposite KLDs in the following way,

$$D_{\bar{KL}}(P \parallel Q) = \frac{1}{2}\left[D_{KL}(P \parallel Q) + D_{KL}(Q \parallel P)\right] \cdot \tag{3.7}$$

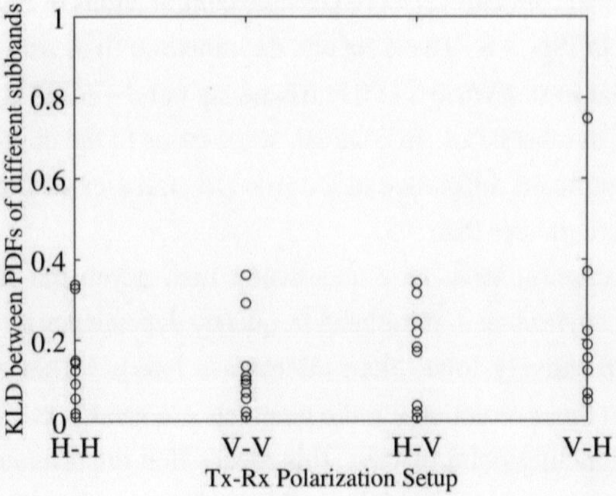

(a). 34 GHz frequency band

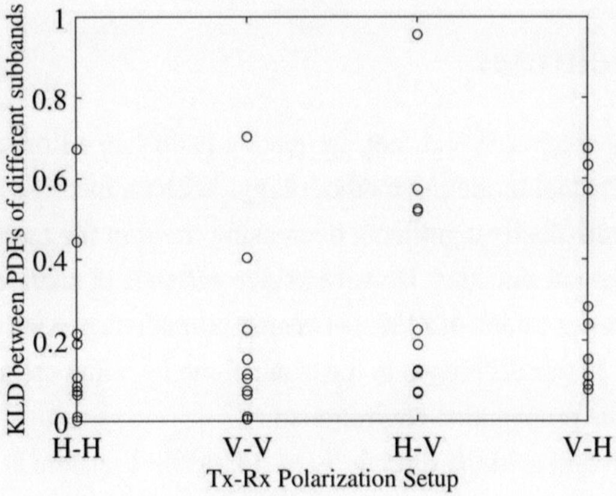

(b). 7 GHz frequency band

Fig. 3.8. KLD between the PDFs of rms DS statistics of sub-bands.

A convergence of KLD in (3.7) towards zero demonstrates that the probability mass functions P and Q are in close agreement with each other. We assume that, if $D_{\bar{KL}}(P \parallel Q) < 0.5$ then difference between the statistics shown by P and Q is not significant. Since whole channel bandwidth is divided into 5 sub-bands, we can compare $\binom{5}{2} = 10$ possible combinations of PDFs, as shown by the circles in Fig. 3.8. These results demonstrate that, rms DS statistics are quite similar in mmWave 34 GHz frequency band—KLD is significantly lower than 0.5 in most cases. In contrast, some cases in the microwave 7GHz band show significant difference in the rms DS statistics as shown by KLD values which are greater than 0.5.

Above discussion leads to a conclusion that, given the same antenna directivity and environment, intra-band frequency dependency in the mmWave bands is comparatively lower than microwave bands. This explains that, channel taps in these directional radio channels are weakly correlated. From the channel modeling point of view, this means that uncorrelated scattering assumption remains valid in modeling of these directional radio channels.

3.5 Conclusion

Studies in this chapter reveal that, frequency itself has a limited impact on defining the channel modeling methodology. With an increase in frequency, there exist a statistically significant decreasing trend in the multipath spatio-temporal dispersion statistics. However, since variance of channel parameters is large, one cannot establish a link between multipath dispersion statistics and the frequency. Major difference in the channel model parameters arise from a difference in the propagation environment.

Bandwidth and antenna directivity of a mmWave system is supposed to be large. Therefore, frequency dependency of sub-bands channels have been analyzed for directional radio channels. Results show that, sub-band rms delay spread statistics of mmWave band do not differ significantly. In other words, channel taps are weakly correlated and uncorrelated scattering assumption can be applied in modeling of UWB mmWave directional radio channels.

Chapter 4

Impact of System Bandwidth on Radio Channel

4.1 Introduction

Using realistic channel measurements, this chapter aims to provide insights into properties of the complex CIR vector when a communication link between Tx and Rx is established only through a single multipath cluster. This scenario is well expected for 5G-NR systems, where a mmWave frequency communication link between a BS and MS is established only through a single strongest Tx-Rx beam pair pointed towards a certain direction in space [8]. A BS serves each MS at a different time slot using time-division multiplexing. The effective single-user radio channel, in such a case, is thus reduced either to a LOS path or to a single multipath cluster. Above discussion reveals that, understanding propagation properties and accurate multipath modeling of each cluster becomes a question of fundamental interest in the mmWave channel modelling research. This depends on how a system illuminates the cluster in terms of bandwidth and antenna directivity, as demonstrated earlier in Section 2.3. Channel models should thus be capable of translating these effects.

Chapter is outlined as follows: Fading of the Rx signal magnitude as a function of system bandwidth is discussed in Section 4.3. From a system design perspective, a large variation in the Rx signal magnitude leads an increased level crossing rate and higher outage probabilities [37]. These quantities have implications on the beam handover protocols and frame structure design, which are required to maintain a communication link. Primary focus of Section 4.4 is to describe randomness in the cross polarization power ratio (XPR) as a function of bandwidth. This basically defines randomness in the 2×2 polarization coupling matrix shown earlier in Eq. (2.3). When bandwidth and antenna directivity of a system is high, results in Section 4.6 show that, channel impulse response corresponding to intra- cluster MPCs cannot be modelled as a complex Gaussian random vector. Intuitively, this also indicates that, unlike state-of-the-art channel models for mmWave systems, each tap in the CIR is not a complex Gaussian random RV. Finally, the chapter is summarized in Section 4.7.

4.2 Measurement Campaign and Data Processing

Dual-band double directional small-scale fading channel measurements are performed for millimeter wave (30.4 to 37.1 GHz) and FCC-UWB (3.4 to 10.1 GHz) bands. Measurements are carried out with a channel sounder which transmits a chip sequence of length 4095 with a chip rate of 6.75 GHz. Details about the channel sounding equipment are described in [94]. Channel sounder offers an instantaneous 3-dB absolute bandwidth of 4 GHz after calibration, where each frequency domain sample roughly corresponds to an absolute bandwidth of 2 MHz. Dynamic range of the channel sounder is up to 70 dB which enables identification of very weak MPCs. Measured CIR is post-processed by estimating a noise floor N_f (dB) using a procedure described in [95] and all time domain samples below $N_f + 10$ dB are zeroed out. Finally, a 25 dB dynamic range is applied to the CIR.

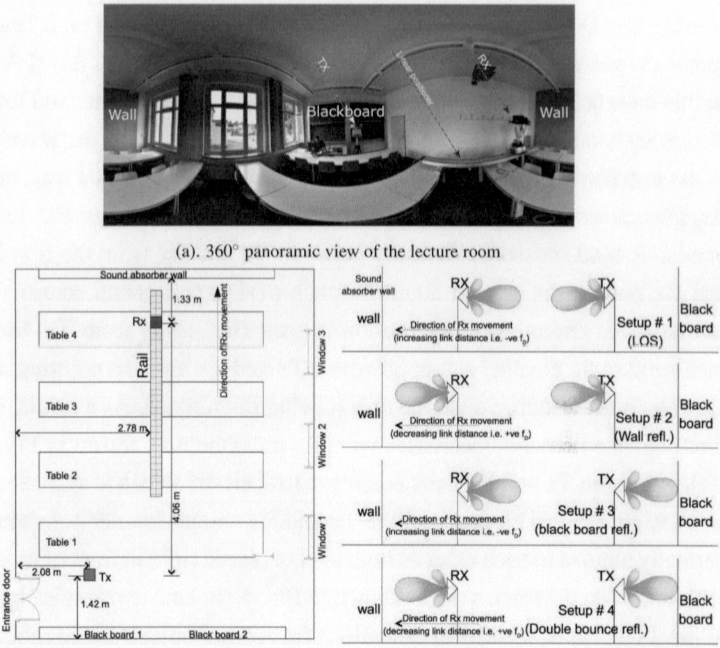

(a). 360° panoramic view of the lecture room

(b) (c)

Fig. 4.1. Measurement scenario of small lecture room H-1519 at TU Ilmenau campus and double-directional measurement setups (a) Overview of Tx-Rx positions, (b) Overview of Tx-Rx antenna pointing angles, f_D corresponds to the Doppler frequency.

Measurements are conducted in a small lecture room shown in Fig. 4.1a. One of the inner brick made walls is covered by the metallic blackboards and the opposing wall is covered with 5 cm thick sound absorbers, whereas the outer wall has metallic frame windows. Doors and windows of the room are closed during measurements to ensure a static time-invariant propagation environment. The Tx is placed at a fixed position near the blackboard, and the Rx unit is placed on a rail with an initial 4.06 meter distance between Tx and Rx, as shown in Figs. 4.1b and 4.1c. In order to track changes in the time-varying radio channel, the Rx is moved in a certain direction with a step distance of Δx_s according to the setups shown in Fig. 4.1c. To measure an aliasing free Doppler power spectrum, Rx step distance should be at least two snapshots per wavelength λ_c of the center frequency f_c — i.e., $\Delta x_s \leq \frac{\lambda_c}{2}$ [96]. In this measurement campaign, the value $\Delta x_s = 2$ mm is maintained for both frequency bands which implies measuring 4 and 21 snapshots per wavelength at the mmWave and the FCC-UWB band, respectively. In this way, chosen Δx_s guarantees aliasing free Doppler power spectrum measurements for both bands. A total measured distance between Tx and Rx from the first to the last Rx position is around 30 cm in each of 4 experimental setups shown in Fig. 4.1c. Direction of Rx motion is same (i.e., away from Tx) for each measured radio channel setup; however, Tx and Rx antenna pointing angles may change depending upon the object being illuminated. As a result, Tx-Rx link distance increases/decreases with Rx movement as shown in Fig. 4.1c. Height of both Tx and Rx units is around 1.68 meters which is approximately an eyesight height. Notice that, the Tx and Rx antenna center points are not perfectly aligned to each other as both are not placed right in front of or behind each other (on the same axis) as shown in Fig. 4.1b. Due to this misalignment in the Tx-Rx axis, additional reflections from other objects in the room such as chairs and tables are also possible along with reflections from intended surface.

For both FCC and mmWave bands, dual polarized Tx-Rx horn antennas each with a 3-dB half power beamwidth (HPBW) of 30° are used. These antennas are used as a compromise between low and high directive antennas due to the following reasons:

1. Decreasing antenna directivity, e.g., to omni-directional antennas may result in MPC contributions from a wide angular spread, thus violating the cluster definition[1]. Therefore, additional filtering in the delay and space domain would be required.

2. Very high gain antennas, in an extreme case, may illuminate only a single MPC and not a group of MPCs to introduce small scale fading. In

[1]We follow the most general definition of a multipath cluster—i.e., a group of MPCs propagating together in delay and space domains having similar departure and arrival angles.

Table 4.1. Channel sounding and measurement parameters

Parameters	Values
Center frequency (f_c)	7 and 34 GHz
Average antenna XPD	30-35 dB
Measured bandwidth	3 dB absolute bandwidth of 4 GHz
Antenna HPBW	30°
Antenna gain	11-12 dBi
Initial Tx-Rx distance	4.06 m
Measured distance on the rail	30 cm
Rx movement step size	2 mm

such a case, if a single MPC is illuminated, there would be no random process to model.

During the measurements, different reflection surfaces are illuminated such that the intersection area of 3-dB Tx-Rx antenna footprints is around 3.4 square meters on the surface. This intersection area imitate a multipath cluster referred to as geometry-based cluster in literature [97]. Geometry-based clusters imitate a physical interacting object (IO) for MPCs in a realistic environment. On the other hand, conventional parameter-based clusters are based on the similarity of multipath channel parameters such as their time-delay, departure and arrival angles. Parameter-based clusters do not necessarily correspond to a particular IO in an environment and much of their accuracy depends upon the applied multipath clustering algorithm. Furthermore, an extraction of parameter-based clusters require array measurements, which is not the case here.

4.3 Fading of the Receive Signal Magnitude

Primary objective of this section is to demonstrate impact of different reflection surfaces and fading depth or standard deviation of the Rx signal magnitude as function of bandwidth. Intuitively, when intra-cluster MPCs are resolved with an increase in bandwidth, fading depth of the Rx signal magnitude is expected to be reduced [42]. In our analysis of Rx signal fading depth, the channel is normalized such that its mean value remains constants; irrespective of the measurement scenario and channel bandwidth.

Mean Value.

Normalization of the CTF $H_W^{ab}(t,f)$ is done such that the ensemble average of the time-varying Rx signal magnitude $m_{\mathrm{Rx}}(t)$ is

$$\rho = E\{m_{\mathrm{Rx}}(t)\} = 1 \tag{4.1}$$

where, from the Parseval's relation

$$m_{\mathrm{Rx}}(t) = \sqrt{\int_0^{\infty} \left|h_W^{ab}(t,\tau)\right|^2 d\tau} = \sqrt{\int_{-\infty}^{\infty} \left|H_W^{ab}(t,f)\right|^2 df}\cdot \tag{4.2}$$

This normalization removes the global pathloss (due to an initial Tx-Rx distance) in each measurement. However, the local pathloss or path gain effect due to the movement afterwards on the rail is maintained which introduces a Doppler shift in MPCs. Additionally, the normalization in (4.1) implies that the mean value of the $m_{\mathrm{Rx}}(t)$ remains constant, irrespective of the bandwidth and the measurement setup.

Standard Deviation (σ_0).

Now we use standard deviation to characterize variations in the Rx signal magnitude. Results in Fig. 4.2b compare σ_0 as a function of W. It can be observed that σ_0 differs considerably in different propagation setups implying a strong influence of scattered MPCs in defining the cluster fading behavior.

Intuitively, depending upon a propagation scenario, σ_0 should decrease upto a particular bandwidth W_0 known as stability bandwidth[2] [42]. An increase in bandwidth beyond W_0 does not result in a considerable change in Rx signal fading statistics and the value of σ_0. Therefore, a dual slope regression fit is quite intuitive and it results in a reasonably good approximation. Higher order polynomials may provide a better approximation, but the slope-constant model in (4.3) is proposed to maintain simplicity and intuitiveness,

$$\sigma_0^{ab}(W) = \begin{cases} c_0 - m\log_{10} W, & \text{if } W \leq W_0 \\ c_1 & \text{otherwise}\cdot \end{cases} \tag{4.3}$$

In (4.3), the bandwidth W is in GHz, the slope m describes the rate of descent in $\sigma_0^{ab}(W)$, constant c_0 is the average amplitude of inphase or quadrature components of plane waves in narrow-band channels and c_1 is their amplitude at the stability bandwidth W_0. Table 4.2 shows parameter values of the model in (4.3) for co- and cross polarized measured channels. It is interesting to note that on average c_0 has higher values in case of wall reflections, demonstrating a large amplitude fading in narrow-band channels. However, the slope factor m is also higher than other propagation setups, demonstrating a relatively faster convergence towards non-fading regime with an increase in W. The reason behind this observation is that MPCs are relatively well separated in the delay domain in case of wall reflections than other measured channel setups. Consequently, intra-cluster MPCs are resolved more quickly with W resulting in higher values of m. The mean square error (MSE) between measurements and the model in (4.3) is calculated as

$$\xi = \sum_{i=1}^{n_B} \left(\underbrace{\sigma_0^{ab}(W_i)}_{\text{model in (4.3)}} - \underbrace{\tilde{\sigma}_0^{ab}(W_i)}_{\text{est. from measurements}} \right)^2 \tag{4.4}$$

[2] *Stability bandwidth* corresponds the system bandwidth at which most of MPCs in the channel are resolved. Therefore, an increase in bandwidth beyond W_0 does not cause any significant variations in the fading envelope.

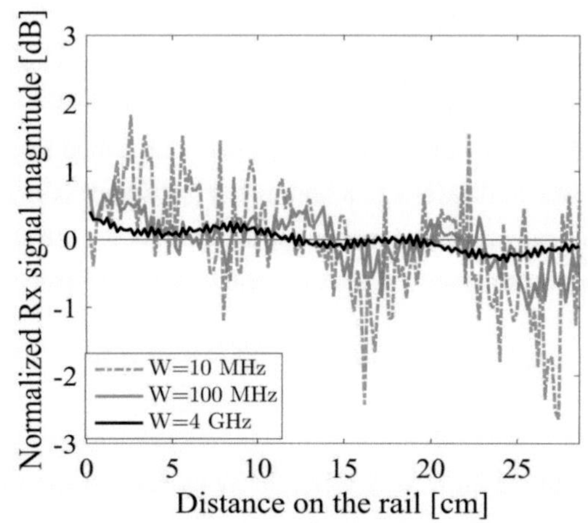

(a)

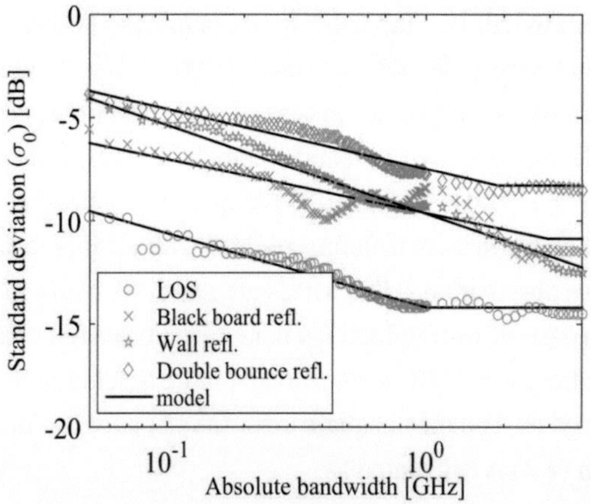

(b)

Fig. 4.2. (a) Fading behavior comparison for different channel bandwidths, setup #1 (LOS), V-V polarization, (b) Standard deviation (σ_0) of the received signal magnitude as a function of channel absolute bandwidth, mmWave band, V-V polarization.

where, n_B corresponds to a total number of frequency samples. Table 4.2 shows the normalized MSE (ξ_n), computed by normalizing ξ in (4.4) by

Table 4.2. Parameter values for empirical model of average diffuse MPCs amplitudes as a function of bandwidth– mmWave band

		LOS	Black board refl.	wall refl.	double bounce refl. (Wall and black board)
VV	m	3.6	2.7	4.3	2.9
	c_0	18	15	29	18.6
	c_1	-14.2	-11.3	-12.2	-8.3
	W_0	1 GHz	3 GHz	4 GHz	2 GHz
	$\log_{10}(\xi_n)$	-3.2	-2.3	-2.8	-2.5
HH	m	2.8	1.7	3.2	1.7
	c_0	12.3	7.1	18.2	8
	c_1	-12.7	-9.5	-10.5	-8.4
	W_0	1 GHz	1 GHz	1 GHz	1 GHz
	$\log_{10}(\xi_n)$	-2.5	-2.2	-1.8	-2
VH	m	2.5	2	3.5	3.1
	c_0	14.3	12.2	22.4	21.1
	c_1	-9.2	-6.7	-11	-8.8
	W_0	2 GHz	2 GHz	4 GHz	2 GHz
	$\log_{10}(\xi_n)$	-2	-1.4	-2.3	-3

$\sum_{i=1}^{n_B} \tilde{\sigma}_0^{ab}(W_i)$. Very small values of the MSE in the Table 4.2 explain a goodness in agreement between measurements and the model in (4.3).

4.4 Cross Polarization Power Ratio (XPR)

Ideally, polarization of a received signal must exactly be the same as that of a transmitted signal to capture maximum of the signal energy. However, a wireless channel and antennas rotate and thus depolarize a transmitted signal which causes a power leakage into the orthogonal signal polarization. Due to this depolarization, in NLOS scenarios, the 2×2 polarization coupling matrix shown in Eq. (2.3) does not remain an identity matrix. The degree of a signal depolarization is known to be a function of carrier frequency and reflection coefficient of the interacting surface [98].

In narrow-band channels, minuscule Tx/Rx movements change the delay and phase relationship of MPCs in the PDPs, causing a random amplitude fluctuation of resolvable delay bins. As a result, total channel power of co- and cross polarized propagation channels may vary significantly with small scale movements. To model this effect, XPR is usually assumed to be a log-normally distributed random variable [11, 99]. This, in turn, causes a randomness in the polarimetric scattering matrix as well. When bandwidth of a system increases, it reduces fading of the Rx signal magnitude and total power of co- and cross polarized propagation channels. Consequently, randomness in the polarization coupling matrix is expected to be reduced. From channel modeling perspectives, XPR power level in a particular propagation environment and its degree of randomness are of fundamental interest.

In this section, XPR behavior is characterized as a function of bandwidth for all 4 different propagation setups shown in Fig. 4.1c. It has been observed that in all cases, the coefficient of variation of XPR around its mean value reduces exponentially with an increase in channel bandwidth demonstrating an evolution towards quasi-deterministic channel fading behaviour. Additionally, a 5-10 dB difference in XPR values has also been observed between narrow-band and UWB channels in the considered propagation setups.

XPR Computation.

The XPR channel power $\mathscr{X}_W^a(x)$ for a particular bandwidth W is computed as

$$\mathscr{X}_W^a(x) = \frac{\int_{f_l}^{f_h} \left|H_W^{aa}(x,f)\right|^2 df}{\int_{f_l}^{f_h} \left|H_W^{ab}(x,f)\right|^2 df} \tag{4.5}$$

where, $H_W^{aa}(x,f)$ corresponds to a channel when a signal transmitted with polarization a is received with same polarization.

Randomness in the XPR.

Degree of randomness in XPR is quantified by coefficient of variation (CV), which is the ratio of standard deviation of $\mathscr{X}_W^a(x)$ at all positions to their ensemble value.

$$\mathrm{CV}_a(W) = 100\frac{\mathrm{std}_x\left\{\mathscr{X}_W^a(x)\right\}}{\mathscr{E}_x\left\{\mathscr{X}_W^a(x)\right\}} \tag{4.6}$$

Results.

By definition, XPR$\rightarrow\infty$ refers to a situation where a signal's polarization is perfectly maintained, XPR$\rightarrow$ 0 dB corresponds to a perfect signal depolarization and XPR$<$ 0 implies large distortion with orthogonal polarization. XPR power level in dB for horizontal to vertical ($\mathrm{XPR_H}$) is defined as the channel power ratio between H-H (Tx horizontal-Rx horizontal) and H-V polarized channels whereby vertical to horizontal ($\mathrm{XPR_V}$) is the power ratio between V-V and V-H setups. Figs. 4.3a and 4.3b show average values of $\mathrm{XPR_H}$ and $\mathrm{XPR_V}$ computed over a small scale area. As expected, XPR power is higher in LOS, demonstrating that polarization is maintained significantly. Theoretically, in the LOS measurements, XPR power should be infinite if the channel is composed only of a single direct LOS path and antenna polarizations are perfectly decoupled. However, in our measurements, in addition to antenna XPDs, MPCs reflected from tables/chairs are also present in LOS channel. Depolarization of these reflected MPCs cause a finite but relatively higher XPR power as compared to other propagation setups. Due to relatively smooth surface of the black board as compared to wall, $\mathrm{XPR_H}$ and $\mathrm{XPR_V}$ values are higher in case of reflection from black board. This demonstrates that depolarization of a transmitted signal is relatively lower for black board reflections as compared to wall which is theoretically expected. Interestingly, double reflections (reflected first from black board and then sound absorber wall) also result in higher XPR power as compared to reflections from wall. This is indeed possible, as polarization rotations by multiple reflections may

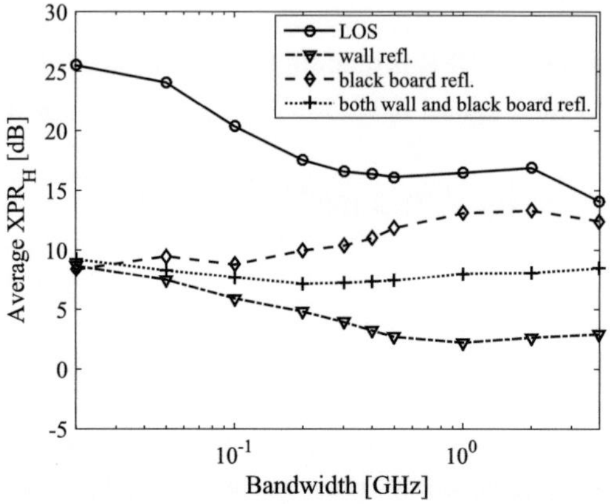

(a)

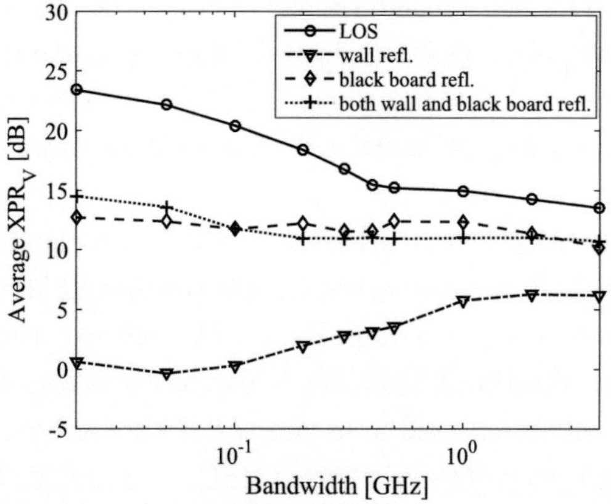

(b)

Fig. 4.3. Average XPR for (a) horizontal to vertical XPR_H (b) vertical to horizontal XPR_V versus channel bandwidth.

end up in a signal orientation parallel to Rx antenna polarization. Results in Figs. 4.3a and 4.3b do not show any conclusive relationship of average XPR values with bandwidth. However, in certain cases a 5-10 dB difference in narrow-band and UWB channels is apparent.

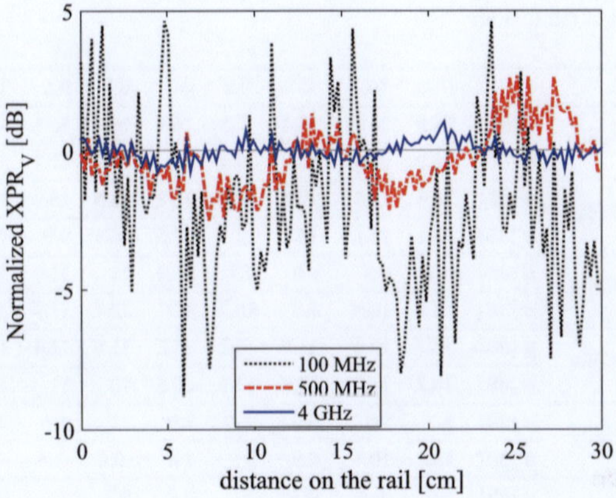

Fig. 4.4. XPR behavior comparison for different channel bandwidths.

Now we compare difference in the degree-of-randomness of XPR values for narrow-band and UWB channels. In Fig. 4.4, XPR values (normalized to 0-dB mean) are shown for each position on the 30cm long measurement track. These results clearly indicate that, an increase in bandwidth reduces variation in the XPR power. Results in Fig. 4.5 show the co-efficient of variation (in %) defined in Eq. (4.6) as a function of bandwidth. It can easily be followed that both CV_H and CV_V converge towards zero percent with an increase in channel bandwidth. This convergence is almost linear along the log-bandwidth scale (i.e. exponential in linear bandwidth scale) in many cases. This clearly demonstrates that XPR behavior becomes more and more deterministic with an increase in channel bandwidth. If only a single direct LOS path is present in the channel, standard deviation in the XPR power must be zero for each channel bandwidth. However, it is not the case in our measurements due to considerably wider opening angle (30° HPBW) of antennas and 25 dB dynamic threshold. Results in Figs. 4.5a and 4.5b show that coefficient of variation for LOS case is lowest in most of the bandwidth setups. However, in our measurements, no clear relationship is observed between the degree of

Table 4.3. Average XPR (μ) and standard deviation (σ) values for the considered propagation setups as a function of bandwidth

Bandwidth [GHz]			0.02	0.05	0.1	0.2	0.3	0.4	0.5	1	2	4
LOS	XPR_H	μ [dB]	25.5	24.05	20.4	17.5	16.6	16.4	16.1	16.5	16.9	14
		σ [dB]	25.9	23.8	18.9	14.7	14	13.8	12.7	13.1	10.3	6.2
	XPR_V	μ [dB]	23.4	22.1	20.4	18.4	16.8	15.5	15.2	15	14	13.5
		σ [dB]	26.1	23.1	19.2	15.3	12.5	10.3	9.6	7.2	6.4	2.5
Black board refl.	XPR_H	μ [dB]	8.4	9.5	8.8	9.9	10.4	11	11.9	13.3	13.5	12.4
		σ [dB]	12.1	12.8	8.6	10.7	9.7	10.4	11.9	10.9	7.9	6.4
	XPR_V	μ [dB]	12.7	12.4	11.7	12.2	11.5	11.5	12.4	12.3	11	10.2
		σ [dB]	16.8	15.9	13.8	12.1	10.5	10	11	10.8	7.8	6.3
wall refl.	XPR_H	μ [dB]	8.7	7.5	5.9	4.8	3.9	3.2	2.7	2.3	2.9	2.9
		σ [dB]	13.1	10.4	6.5	3	1.4	-0.6	-1.5	-1.4	-0.9	-1.9
	XPR_V	μ [dB]	0.6	-0.4	0.3	2	2.8	3.2	3.6	6	6.9	6.2
		σ [dB]	1.2	-0.9	-1.2	-0.2	-0.2	-1	-1.4	0.2	0.3	-1.1
double bounce refl.	XPR_H	μ [dB]	9.2	8.3	7.7	7.1	7.2	7.3	7.4	8.2	8.4	8.5
		σ [dB]	11.2	9.7	8.1	6.4	6.1	5.8	5.8	5.3	5.6	5.2
	XPR_V	μ [dB]	14.5	13.6	11.8	11	10.9	11	10.9	11.1	11.2	10.8
		σ [dB]	22.9	21.7	18.2	15.4	14.3	13.5	12.8	10.1	10.1	9.4

randomness of XPR and different reflecting surfaces. Table 4.3 summarizes mean and standard deviation values of XPR_H and XPR_V for the experimental setups shown in Fig. 4.1c. Note that standard deviation of XPR also shows a decreasing trend with an increase in the channel bandwidth.

4.5 Validity of Uncorrelated Scattering Assumption

By definition, uncorrelated scattering means that the energy falling into each delay tap comes from an independent scatterer. When only a single cluster is illuminated in a propagation channel, then in the narrowband channels, uncorrelated scattering assumption becomes hard to be justified due to increased smearing of intra-cluster MPCs in the delay domain. However, with an increase in bandwidth, intra-cluster MPCs are resolved in the delay do-

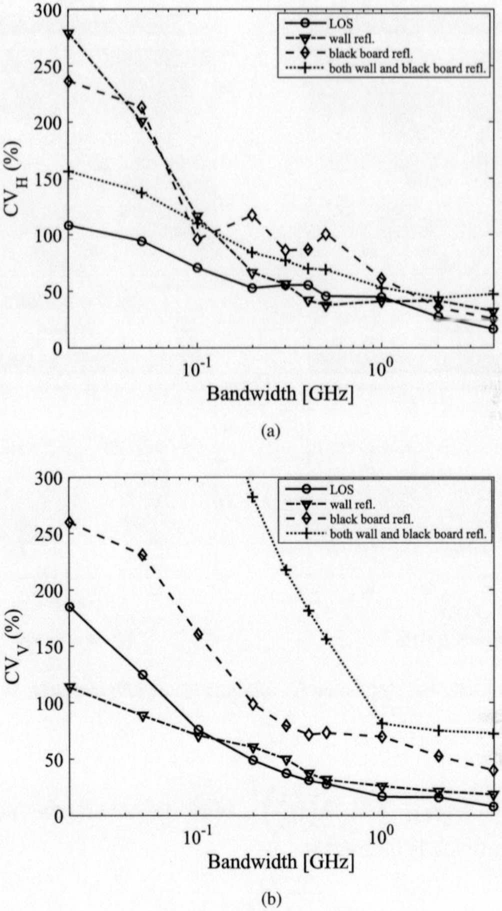

Fig. 4.5. Co-efficient of variation for (a) horizontal to vertical (CV_H) (b) vertical to horizontal (CV_V) versus channel bandwidth.

main and energy falling into each tap may only come from a single intra-cluster MPC. This leads towards uncorrelated scattering. Assuming that, L be the number of intra-cluster channel taps in the delay domain and each tap arrives at the delay time τ_i then the CIR vector at time t can be expressed

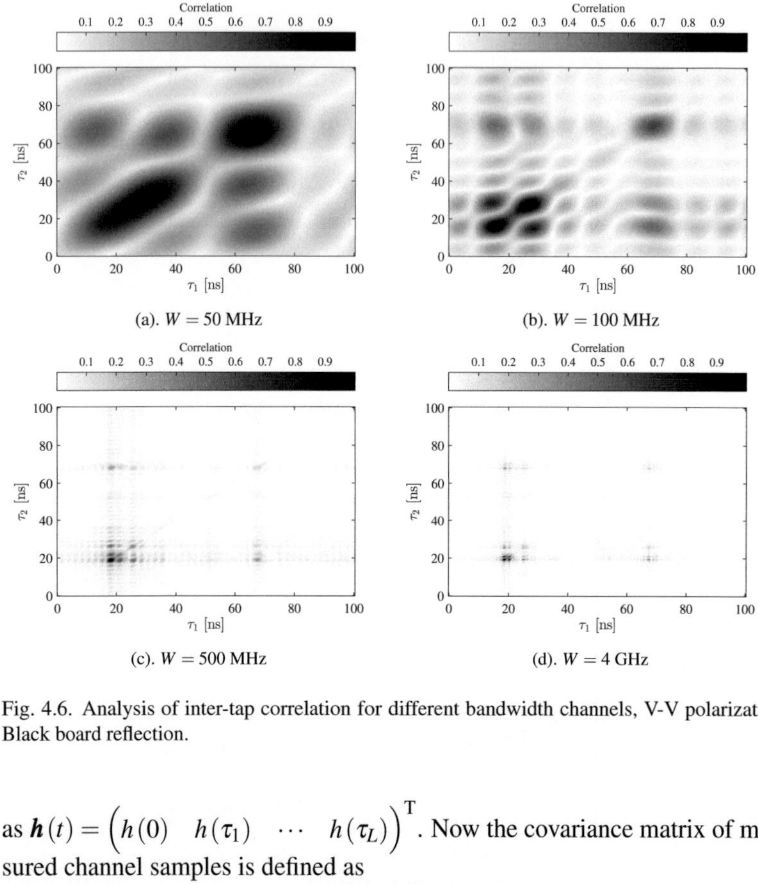

Fig. 4.6. Analysis of inter-tap correlation for different bandwidth channels, V-V polarization, Black board reflection.

as $\boldsymbol{h}(t) = \begin{pmatrix} h(0) & h(\tau_1) & \cdots & h(\tau_L) \end{pmatrix}^{\mathrm{T}}$. Now the covariance matrix of measured channel samples is defined as

$$\boldsymbol{K}_s = E\left\{\boldsymbol{h}\boldsymbol{h}^H\right\}. \tag{4.7}$$

In order to demonstrate inter tap correlation, results in Fig. 4.6 compare absolute values of the covariance matrix $\boldsymbol{K}_s$ for different bandwidth channels. These results match perfectly with intuition and one can clearly see that correlation between adjacent channel taps reduces significantly with an increase in channel bandwidth.

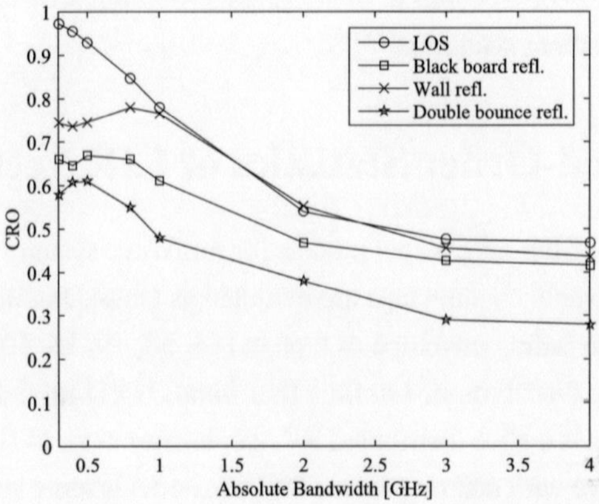

Fig. 4.7. Close to rank one approximation as a function of bandwidth, V-V polarization.

An increase in the number of significant eigen values of the covariance matrix $\boldsymbol{K}_s$ characterizes the number of stochastic degrees-of-freedom in the frequency domain [41]. In other words, it means that, inter-tap correlation in the complex CIR reduces. In order to characterize de-correlation between the channel taps, we analyze closeness to rank one (CRO) of the covariance matrix $\boldsymbol{K}_s$. Let $\lambda_1 \geq \lambda_2 \geq \lambda_3 \cdots \lambda_L$ be the eigen values of covariance matrix $\boldsymbol{K}_s$, then CRO can be calculated as

$$\text{CRO} = \frac{\lambda_1}{\sum_{i=1}^{L} \lambda_i} \tag{4.8}$$

where, $0 \leq \text{CRO} \leq 1$. By definition, CRO $\rightarrow 0$ refers to a situation when MPCs are perfectly de-correlated. On the other hand, CRO $\rightarrow 1$ corresponds to the case when MPCs are highly correlated. In Fig. 4.7 CRO values are shown as a function of bandwidth for different measurements setups shown in Fig. 4.1. It can be observed that intra-cluster MPCs de-correlate with the bandwidth. Notice that, CRO values in case of double bounce reflection are lowest as compared to other measured channels. This implies that, the number of stochastic degrees-of-freedom in the frequency-domain are higher compared to other measurement scenarios.

4.6 Second-Order Statistics of CIR Vector

Review of state-of-the-art channel models for mmWave systems in Section 2.6 shows that resolvable channel taps are modeled as Gaussian random variables. For example, the fading envelope of taps in [14, 58, 19, 15, 40, 18, 3] follow Rayleigh fading distribution. On the other hand, NYU model assumes that each channel tap is a Rice distributed RV. Zöchmann et al. [47] demonstrated that the Two-wave with diffuse power (TWDP) model is more suitable than the Rician fading model. Major difference between the Rice and TWDP fading model is that the former assumes single fixed amplitude MPC, whereas the later assumes two MPCs. Romero et al. [48] proposed Fluctuating two ray (FTR) model based on the experimental data in [72]. Recalling from Section 2.4 that, FTR model generalizes the TWDP model by introducing random fluctuations in both fixed amplitude MPCs. Similar to Rayleigh/Rice fading taps, both TWDP and FTR models represent rich scattering environment due to the circularly symmetric complex Gaussian assumption on the scattered MPCs.

Channel models depend upon system assumptions. Given the same propagation channel, systems with different properties – e.g., bandwidth and beamforming gain – may see a channel differently. Review ofmmWave system properties in Section 1.1 reveal that, in contrast to the 3GPP-LTE systems [9], mmWave systems are supposed to operate with high antenna directivity and large channel bandwidths. High gain antenna act as spatial multipath filter for a propagation channel. High bandwidth reduces smearing of MPCs in PDPs. Consequently, the number MPCs falling into a resolvable delay bin reduces. The central limit theorem, therefore, does not remain valid and channel tap cannot be modeled as a Gaussian RV [13]. Therefore, theoretically, one can expect that the complex CIR does not remain a Gaussian random vector.

Primary motivation of this section is to investigate if the cluster CIR can be modelled as a Gaussian random vector or not? Schuster et. al. in [41], investigated the same question for omni-directional radio channels using indoor 2-5 GHz UWB channel measurements. Akaike Information Criterion (AIC) approach adopted in [41] computed Kullback-Leibler distance between the fading distribution of measured channel taps and well known models from literature including Rayleigh, Rice, Nakagami, log-normal and Weibul families. Results in [41] demonstrated that taps amplitude fading in UWB channels can still be well described Rayleigh and Rice fading distribution. Thus the baseband equivalent CIR can be modelled as a Gaussian random vector. In this section, a new approach is introduced which is based on the analysis of second order statistical functions which analyse temporal correlation properties of the complex CIR vector. Analysis of results demonstrate that, these properties provide a well expected theoretical insight into the measured radio channels. These results have been published in our work [30] and are summarized below.

4.6.1 Temporal Auto-Correlation Function

Let f_D be a Doppler shift in an MPC received at a time delay τ, then the Doppler-variant impulse response $h_W^{ab}(\tau, f_D)$ is Fourier transform of $h_W^{ab}(\tau,t)$ in the slow-time domain t. Let $h_W^{ab}(\tau,t)$ be wide-sense stationary (WSS) in slow-time t, then the Doppler power spectral density (PSD) is defined as

$$\Phi_{\boldsymbol{h}}(f_D) = \int_0^{\infty} \left| h_W^{ab}(\tau, f_D) \right|^2 d\tau. \tag{4.9}$$

Let $\boldsymbol{h}(t)$ denote a random sample of the measured multipath CIR vector at a time instant t, then the temporal ACF $c_{\boldsymbol{h}}(\Delta t) = E\{\boldsymbol{h}(t)\boldsymbol{h}^*(t+\Delta t)\}$[3] is equivalent to the inverse Fourier transform of $\Phi_{\boldsymbol{h}}(f_D)$. For $\Delta t > 0$, the absolute value of $c_{\boldsymbol{h}}(\Delta t)$ quantifies the channel coherence time (or distance) and its knowledge plays a key role in channel tracking algorithms [100]. For the

[3] $(\cdot)^*$ denotes the complex conjugate of the function.

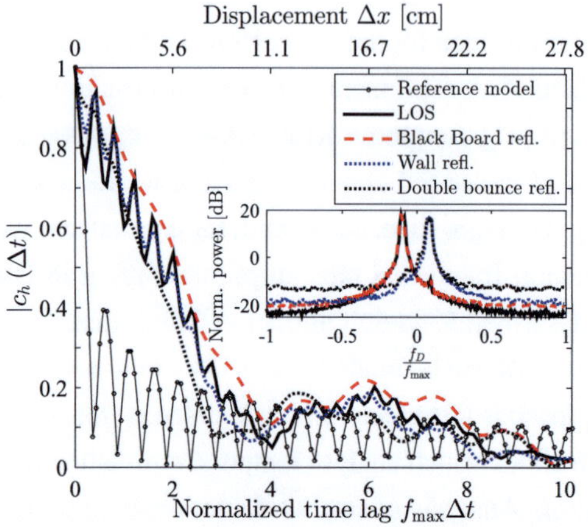

(a). FCC band

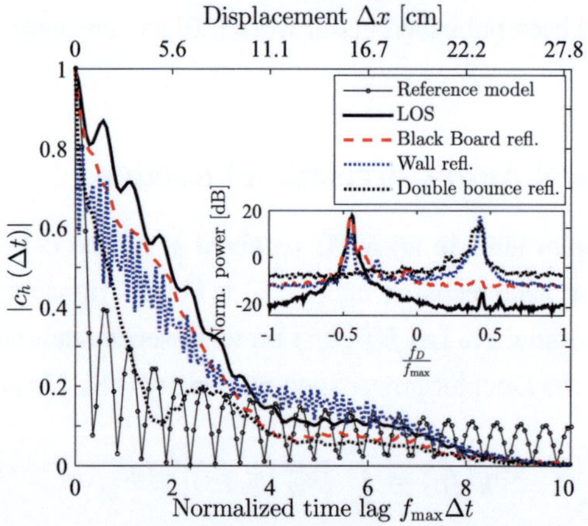

(b). mmWave

Fig. 4.8. Analysis of ACF for different reflection surfaces, V-V polarization, $f_{\max} = 100$ Hz, $W = 4$ GHz. Speed of the Rx is equal to 10 kilometer per hour. Recalling that the measurements do not consider a highly controlled environment such that MPCs are received only from an intended reflection surface. Therefore, in some measurements, an abuse of multipath cluster definition (same delay and the same AoA of MPCs) is possible.

reference *Rayleigh-Rice* multipath model, $c_{\boldsymbol{h}}(\Delta t) = J_0(2\pi f_{\max}\Delta t)$ is a real valued function having a symmetric U-shaped Doppler PSD, where J_0 is a zeroth-order Bessel function of first kind and $f_{\max}$ is the maximum Doppler frequency. However, $c_{\boldsymbol{h}}(\Delta t)$ in realistic channels is expected to be a complex valued function with asymmetric Doppler PSDs which are caused either by non-isotropic scattering from the propagation environment or by the use of directional antennas as shown in Fig. 4.8[4]. Clearly, the ACFs of measured multipath clusters differ considerably from the reference model. In case of FCC band, a high similarity between the Doppler PSDs (and ACFs) explains that the impact of surface scattering is not very significant. For the wall and double bounce reflections, a comparison of Doppler PSDs of both bands shows that number of peaks[5] in the mmWave band are higher than FCC band. These observations are in-line with the hypothesis in [35], which states that smaller wavelengths at mmWave frequencies may lead to more diffuse scattering than lower frequency bands. Assuming that a channel is correlated if $|c_{\boldsymbol{h}}(\Delta t)| \geq 0.5$, results in Table 4.4 summarize correlation distances in measured channels. One may note that, for the mmWave band, correlation distances for different measurement setups differ considerably from each other. On the other hand, they are quite similar for FCC band. This observation clearly demonstrates that, the influence of surface scattering on the correlation distance is more pronounced in the mmWave band than in the FCC band.

4.6.2 Temporal Cross-Correlation and Complementary Auto-Correlation Functions

A Gaussian random vector is the one having uncorrelated inphase and quadrature components. In order to analyze, if a measured CIR vector $\boldsymbol{h}(t)$ is

[4]Note that comparisons shown in Figs. 4.8 and 4.9 consider same absolute bandwidth (W) for both FCC and mmWave frequency bands. Similar comparisons with same relative bandwidth may also provide an interesting insight into the fading statistics. For example, due to increased smearing of MPCs, we expect that fading statistics of clusters at the FCC band would be comparatively more closer to the reference models.

[5]By *peaks*, we mean the number of ripples in the Doppler PSDs shown in Fig. 4.8 which correspond to the intra-cluster scattered MPCs.

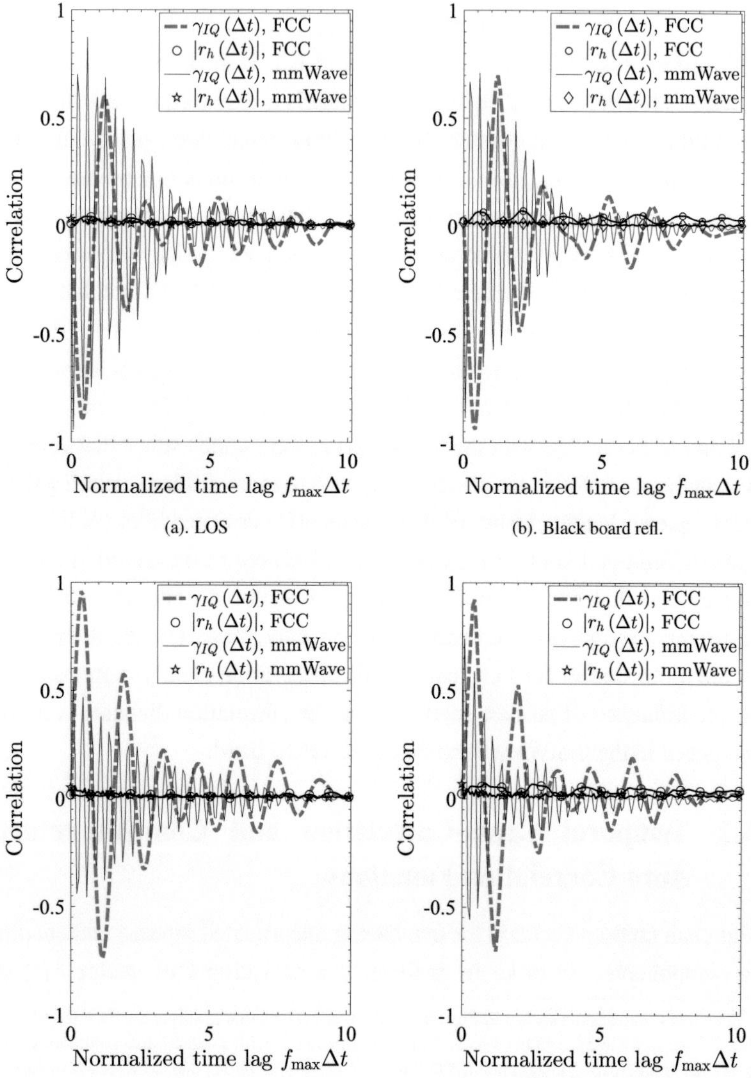

(a). LOS (b). Black board refl.

(c). Wall refl. (d). Double bounce refl.

Fig. 4.9. Analysis of CACF and the CCF for different measurement setups, V-V antenna polarization setup, $f_{\max} = 100$ Hz, $W = 4$ GHz. Speed of the Rx is equal to 10 kilometer per hour.

Table 4.4. Correlation distance in terms of wavelength for different clusters

	LOS	Black board refl.	Wall refl.	Double bounce refl.	Rayleigh/ Rice
FCC band	$1.3\lambda_c$	$1.4\lambda_c$	$1.3\lambda_c$	$1\lambda_c$	$0.1\lambda_c$
mmWave band	$6.1\lambda_c$	$5.5\lambda_c$	$5.2\lambda_c$	$2\lambda_c$	$0.7\lambda_c$

a Gaussian random vector, we analyze temporal cross-correlation function (CCF) $\gamma_{IQ}(t_1,t_2) = E\left\{\Re(\boldsymbol{h}(t_1))\Im\left(\boldsymbol{h}^T(t_2)\right)\right\}$. The temporal CCF measures a correlation between real ($\Re$) and imaginary ($\Im$) parts of CIR vector. Recalling that, a vanishing CCF $\gamma_{IQ}(\Delta t) \approx 0, \forall \Delta t \geq 0$ explains that the CIR is a complex Gaussian random vector which corresponds to a rich scattering channel and vice versa. Additionally, for both FCC and mmWave bands, a comparative analysis is done to investigate if the measured CIR is a circularly symmetric complex random vector or not? This is analyzed by the complementary ACF (CACF), i.e., $r_{\boldsymbol{h}}(t_1,t_2) = E\left\{\boldsymbol{h}(t_1)\boldsymbol{h}^T(t_2)\right\}$, which measures a correlation between $\boldsymbol{h}(t)$ and its complex conjugate $\boldsymbol{h}^*(t)$. A channel $\boldsymbol{h}(t)$ is said to be second-order stationary, if it is WSS and the CACF is only a function of Δt [101]. Propagation channel $\boldsymbol{h}(t)$ with a vanishing CACF $r_{\boldsymbol{h}}(\Delta t) \approx 0, \forall \Delta t \geq 0$ is a circularly symmetric complex random vector [102]. Assuming that $\boldsymbol{h}(t)$ is second-order stationary, then the real and imaginary parts of $r_{\boldsymbol{h}}(\Delta t)$ are defined as

$$\Re\{r_{\boldsymbol{h}}(\Delta t)\} = \mu_{II}(\Delta t) - \mu_{QQ}(\Delta t) \tag{4.10}$$

and

$$\Im\{r_{\boldsymbol{h}}(\Delta t)\} = \gamma_{IQ}(\Delta t) + \gamma_{QI}(\Delta t) \tag{4.11}$$

where, $\mu_{II}(\Delta t)$ and $\mu_{QQ}(\Delta t)$ are the ACFs of I and Q components, respectively. Note that, in case of a circularly symmetric complex Gaussian signal, $\mu_{II}(\Delta t) = \mu_{QQ}(\Delta t)$ and $\gamma_{IQ}(\Delta t) = \gamma_{QI}(\Delta t) = 0$, resulting in $r_{\boldsymbol{h}}(\Delta t) = 0, \forall \Delta t \geq 0$. However, every circularly symmetric signal is not always a complex Gaussian random variate. In fact, $r_{\boldsymbol{h}}(\Delta t) = 0$ also when

$\mu_{II}(\Delta t) = \mu_{QQ}(\Delta t)$ and $\gamma_{IQ}(\Delta t)$ is a non-zero odd function [103]. A non-zero CACF establishes the following facts: *i)* Second-order statistics have not been fully described by $c_{\boldsymbol{h}}(\Delta t)$, *ii)* the channel $\boldsymbol{h}(t)$ is a non-circular or improper complex vector [104]. In this case, beamforming techniques based on the widely linear processing result in higher performance gains over the linear counterparts [104].

Fig. 4.9 shows that both $\gamma_{IQ}(\Delta t)$ and $r_h(\Delta t)$ are roughly equal to zero at $\Delta t = 0$, which explains that I and Q components of $\boldsymbol{h}(t)$ are (instantaneously) mutually uncorrelated. However, $\gamma_{IQ}(\Delta t) \neq 0$ for $\Delta t > 0$ and as such differs from the reference model. For $\Delta t > 0$, Figs. 4.9c and 4.9d show that the values of $\gamma_{IQ}(\Delta t)$ are higher for FCC than the mmWave band. This observation indicates that wall and double bounce clusters when illuminated with FCC band are sparser than mmWave band demonstrating an increased surface scattering at mmWave band. Similarly, for the mmWave band, LOS and black board clusters are sparser than wall and double bounce reflections. This is due to the reduced scattering in LOS and black-board clusters as compared to the wall and double bounce reflections, which agrees with the intuition and from the Doppler PSDs in Fig. 4.8. Non-zero CCFs in Fig. 4.9 imply that the rich scattering assumption made in the Rayleigh, Rice, TWDP and FTR fading models is not satisfied. For all $\Delta t > 0$, the results $\gamma_{IQ}(\Delta t) \neq 0$ and $r_{\boldsymbol{h}}(\Delta t) \approx 0$ demonstrate that $\mu_{II}(\Delta t) = \mu_{QQ}(\Delta t)$ and $\gamma_{IQ}(\Delta t)$ is an odd function. As such, one can conclude that the complex CIR vector $\boldsymbol{h}(t)$ is a circularly symmetric non-Gaussian random vector with a rotationally invariant PDF. These properties are in agreement with correlation properties of sparse channels in [13]. From system performance perspectives, in this case, both linear and its widely linear beamforming counterpart will result in similar performances [104].

Here it is important to emphasize that in narrowband channels, the amplitude of $\gamma_{IQ}(\Delta t)$ decreases but does not vanish to zero due to the asymmetry of the Doppler PSD. The Rx envelope will then converge towards the reference model [27]. Presented results and discussions above lead to the following remark. With limited bandwidth (narrowband assumption) at lower frequency bands, the reference model assumption as used in 3GPP-TR 38.901 [14] and WINNER [11] is valid and sufficient. However, the reference model is not realistic when it comes to, a) larger bandwidth even at bands below 10 GHz, b) mmWave bands with traditionally higher bandwidths.

4.7 Conclusion

Investigations in this chapter show that, in contrast to narrowband channels, fading of taps in mmWave radio channels vanishes with an increase in system bandwidth. Therefore, received signal power does not remain a random process. Additionally, XPR and polarization coupling matrix cannot be modelled as a stochastic random process. Mutual correlation between MPCs reduces with an increase in system bandwidth, so uncorrelated scattering assumption in delay domain can still be applied. Second-order statistics of the complex CIR vector have also been studied. It has been found that rich multipath scatting assumption cannot be justified for an individual cluster and it cannot be modelled as Rayleigh/Rice multipath channel.

Chapter 5

Impact of Antenna Directivity on Radio Channel

5.1 Introduction

In order to mitigate higher path losses at mmWave frequencies, the antenna directivity of mmWave systems is supposed to be considerably high. A propagation channel is, thus illuminated with one or few antenna beams which are pointed towards certain directions in space. Such a selective illumination of the propagation environment may result in a sparse radio channel, as shown in Fig. 5.1. In an extreme case, a highly directive antenna may illuminate the radio channel with only a single MPC. Therefore, during the characterization and modelling of such radio channels, the impact of parameters associated with an MPC such as its amplitude, phase and polarization can significantly be high. This is fundamentally different from narrowband channel models such as WINNER [11], 3GPP-SCM [10] and COST-2100 [12] channel models.

In these models, due to the superposition of a large number of intra-cluster MPCs, propagation properties of individual MPCs lose their signifi-

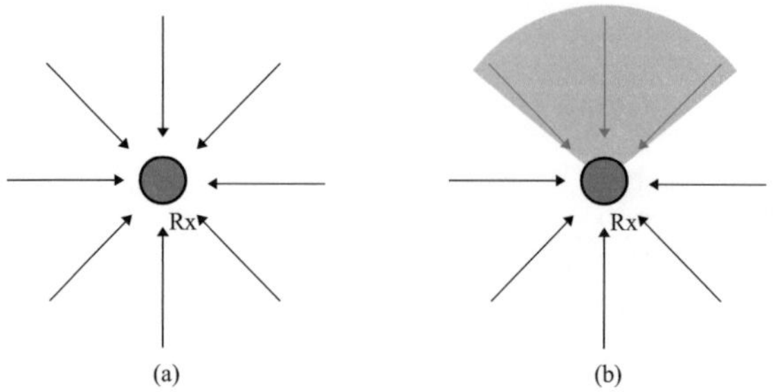

Fig. 5.1. An illustration of how a directive illumination of a propagation channel filters the number of MPCs. (a). Case of Omni-directional antenna. (b). Case of directive antenna.

cance in terms of multipath modelling. Consequently, the multipath modelling approach only aims to model a resolvable channel tap as a complex Gaussian RV without taking care of the actual propagation properties of each individual intra-cluster MPC. Marginal tap statistics thus follow Rayleigh fading distribution and the complex CIR is modelled as Gaussian random vector.

The objective of this chapter is to show the impact of high antenna directivity and large bandwidth of mmWave systems on the small-scale fading statistics. For this purpose, a dedicated measurement campaign is carried out at the 60 GHz frequency band. Radio channel is illuminated with Tx-Rx antennas with different directivities. Details about the measurement campaign are described in Section 5.2. Fade depth reduction with an increase in antenna directivity is demonstrated in Section 5.3. Second-order statistical characterization in Section 5.4 shows that the complex CIR cannot be modelled as a Gaussian random vector when antenna directivity and bandwidth is high. Results in Section 5.5 show that the complex CIR cannot be modelled as a WSS random process in the slow-time domain. However, in the frequency domain, it can still be modelled as a WSS random process. Finally, conclusions are drawn in Section 5.7.

Fig. 5.2. Picture of the environment and fading measurement setup.

5.2 Measurement Campaign

Channel measurements are carried out using ultra-wide band transmit and receive channel sounding units which have been introduced in [77]. After calibration, channel sounder offers an instantaneous 3-dB absolute bandwidth of 4 GHz and a dynamic range up to 70 dB. The sounding signal is a chip sequence of length 4095 with a chiprate of 6.75 GHz. Both Tx and Rx sounding units are synchronized to the same clock via cables resulting in a highly stable relative delay between the sounder units.

The experiments have been conducted in a typical laboratory room with a plenty of equipment placed in the vicinity of both transmitter (Tx) and receiver (Rx), as shown in Fig. 5.2. Position of Tx is kept fix, however, Rx moves towards the wall on a linear positioning device denoted as *rail*, as shown in Fig. 5.3. Both LOS and NLOS[1] measurements are conducted with Tx-Rx antenna setups shown in Table 5.1, which are detailed as follows:

- In the LOS case, omni-directional antennas are used at both Tx and Rx. Small-scale fading measurements are conducted using the experimental setup, rail position and Rx movement direction as shown in Fig. 5.3.

- In the NLOS case, only a particular spot on the wall is illuminated with a 15° half power beam width (HPBW) Tx antenna. Incident point in the wall is 2.6 meters away from the Tx, which is far beyond the Fraunhofer distance (31mm) to ensure plane wave propagation. As a result, Tx

[1] Although, measurements consider a perfect optical visibility between Tx&Rx, the setup is still referred to as NLOS, because directional antennas are not pointed towards each other.

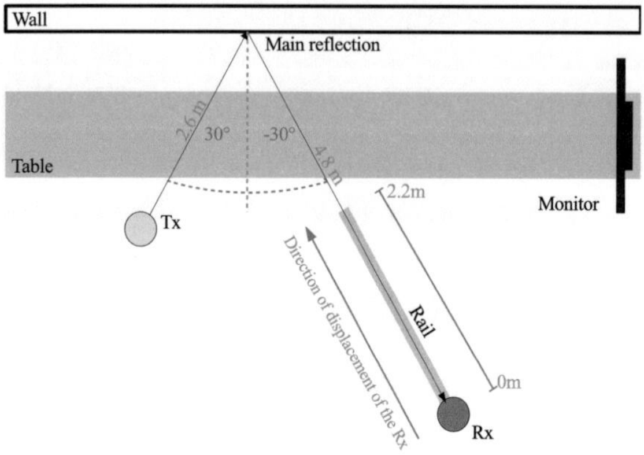

Fig. 5.3. Top-view schematic of the radio channel fading measurement set-up.

antenna footprint on the surface of the wall is approximately 0.4 m^2. At the Rx side, both directional (always pointed towards the incident point) and omni-directional antennas are used. Directional illumination with 15° HPBW Rx antenna, when placed at 4.8m away from the surface of wall, leaves an antenna footprint of 1.25 m^2 which reduces up to 0.4 m^2 when Rx approaches at the terminal point of its movement. Similarly, when a 30° HPBW Rx is employed, the antenna footprint on the surface of wall reduces from 5.2 m^2 up to 0.7 m^2 as Rx moves closer to the wall. Note that, when low gain Rx antennas are employed, a cross-talk between Tx-Rx antennas may result in a possible violation of our NLOS definition. However, we emphasize that, the energy contributions of these cross-talk effects are negligible. Consequently, the measurements with low gain Rx antennas approximately fulfill our NLOS definition. During measurements, Tx and Rx antenna pointing angles are adjusted such that the intersection area of their footprints on the surface of the wall imitate a geometry-based multipath cluster [97].

For both LOS and NLOS measurements, same experimental setup shown in Fig. 5.3 is used, while maintaining the rail position and direction of motion

Table 5.1. Channel sounding and measurement parameters

Parameters	Values
Center frequency (f_c)	60 GHz
Measured bandwidth	3 dB bandwidth of 4 GHz
Tx/Rx height	1.52 m
Rx antenna HPBW	15° (21 dBi) | 30° (14 dBi) | Omni (0 dBi)
Tx antenna HPBW	15° (21 dBi) | Omni (0 dBi)
Measured radio channel setups and their visibility	Tx 15° HPBW, Rx 15° HPBW→NLOS Tx 15° HPBW, Rx 30° HPBW→NLOS Tx 15° HPBW, Rx Omni→NLOS Tx Omni, Rx Omni →LOS
Tx-Rx antenna polarization setup	Vertical-Vertical
Rx movement step size	2.5 mm

of Rx i.e., towards the wall. Note that, high resolution antennas (greater than 15° HPBW) are intentionally avoided during the measurements, as this may lead a situation when only a single MPC is illuminated instead of a cluster. Thus, resulting in a measured channel with a no random process to model. In order to track changes in the time-varying propagation channel, at least two snapshots per wavelength should be measured [96]. Therefore, Rx movement step size is fixed to $\frac{\lambda_c}{2} = 2.5$mm, where, λ_c corresponds to a wavelength at 60 GHz center frequency. In this way, Rx is moved on the rail towards the wall, covering a 2.2m long track in a local area. These linear Rx movements resulted a total of 881 channel snapshots along the track. Three synchronized PCs are used to control linear Rx movements along the rail, switches for the polarization and record the data sent by the UWB sounding units. Table 5.1 summarizes some other important configuration settings used during the measurements.

Channel Data Processing

Let $H^{ab}(t,f)$ be the wide-band channel transfer function (CTF) when a signal is transmitted with a polarization a and received with a polarization b. We assume that the directional CTF is frequency flat, which is quite plausible assumption as directional antennas illuminate only selective objects in the particular directions, hence resulting in very low delay spread values as shown in Fig. 5.4. Let n_f denotes the total number of frequency samples in the CTF at a particular time instant t, then

$$H^{ab}(t,f) = \sum_{n=1}^{n_f} A^{ab}(t,n)\, e^{j\theta^{ab}(t,f)}\, \delta(f - n\Delta f) \tag{5.1}$$

where, A^{ab} and θ^{ab} are magnitude and phase responses of the channel at the nth frequency bin of the CTF at a time instant t. For the bandwidth reduction (if required), frequency domain samples are extracted while maintaining f_c

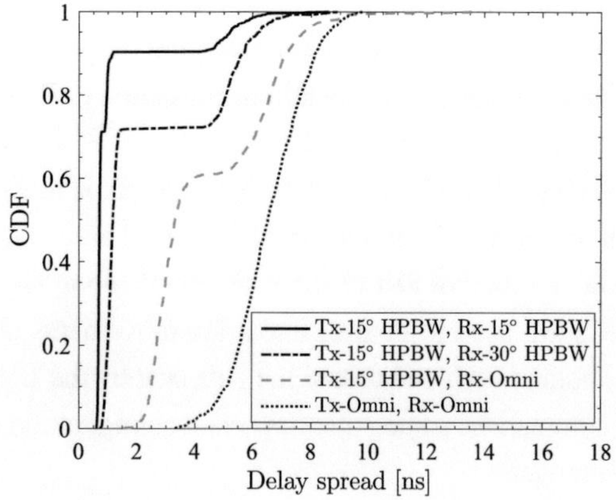

Fig. 5.4. Root mean square (rms) delay spread statistics over the whole measurement track.

and length of the full-band frequency domain channel in following way:

$$H_W^{ab}(t,f) = \begin{cases} H^{ab}(t,f), & f_l \leq f \leq f_h \\ 0 & \text{otherwise.} \end{cases} \quad (5.2)$$

Now the channel absolute bandwidth W is defined as $W = f_h - f_l$, where f_h and f_l are the highest and lowest frequencies of the band, respectively. Let τ be the time delay of an MPC, then the time domain CIR vector $h_W^{ab}(t,\tau)$ is obtained by inverse Fourier transform of $H_W^{ab}(t,f)$.

5.3 Fade Depth Analysis

Fade depth is known to characterize variations in the Rx signal power about its local mean value due to the small-scale fading and it effects the probability of outage [23]. In addition, it is also useful for the convenient evaluation of link-budget analysis. Fade depth analysis provided earlier in Section 4.3 considered only the impact of signal bandwidth while maintaining the antenna directivity at both Tx and Rx. Now we analyse the impact of both mmWave aspects (i.e. antenna directivity and system bandwidth) on variations in the Rx signal magnitude/power $\left(\sqrt{\int_{-\infty}^{\infty} \left|H_W^{ab}(t,f)\right|^2 df}\right)$ which is normalized to 0-dB mean value. Fade depth is then calculated as

$$F = s\sigma_p \quad (5.3)$$

where, σ_p is the standard deviation of power and s corresponds to the desired system performance and outage levels.

Experimental results shown in Fig. 5.5 demonstrate a fade depth reduction with both bandwidth and antenna directivity. Notice that, the measurement track in our experimental setup is quite long (2.2m) and the signal power may not be a stationary random process over the whole track. Therefore, measurement track is divided into approximately stationary intervals each of length 40λ. Fading depth is then computed from different 40λ stationary

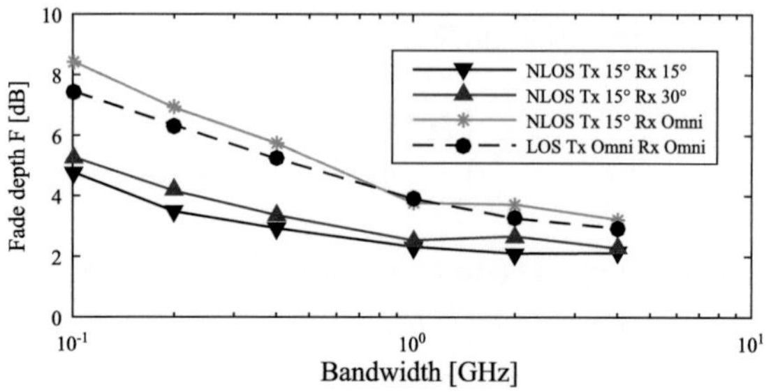

Fig. 5.5. Fade depth analysis of receive signal power as function of antenna directivity and absolute bandwidth of the system, $s = 3$.

Table 5.2. Empirical fade depth model parameters values.

Tx	Rx	Visibility	c_0	m	c_1	α
15° HPBW	Omni	NLOS	6.4	0.67	0.32	1.75
15° HPBW	30° HPBW	NLOS				1.1
15° HPBW	15° HPBW	NLOS				1

intervals which are collected over the measurement track using a sliding window approach. Using the average fading depth values, an empirical slope-constant model is proposed in our work [26]. The model can be written as

$$F(W,s,\alpha) = \begin{cases} s\alpha\left(c_0 - m\log_{10} W\right), & \text{if } W \leq W_0 \\ s\alpha c_1 & \text{otherwise}\cdot \end{cases} \tag{5.4}$$

where, α is the parameter for directivity, c_0 is related to the intercept value, m defines the slope and $s\alpha c_1$ is the model value at the stability bandwidth $W_0 = 1$ GHz. Empirical parameter values of the model in Table 5.2 show that α reduces significantly with an increase in Rx antenna directivity and fade depth reduces accordingly. This explains that, antenna directivity has a significant role in defining the fading depth of Rx signal. The model Eq. (5.4) also explains that, fade depth reduces exponentially with bandwidth of the system.

5.4 Second-Order Statistics of CIR Vector

Temporal Cross-Correlation Function (CCF)

Let $\boldsymbol{h}(t)$ be the $N-$taps complex CIR vector at a time instant t. Let $\Re(\boldsymbol{h}(t))$ and $\Im(\boldsymbol{h}(t))$ represent the real and imaginary parts of $\boldsymbol{h}(t)$ respectively, then their temporal CCF is then defined as

$$\gamma_{IQ}(t_1,t_2) = E\left\{\Re(\boldsymbol{h}(t_1))\Im\left(\boldsymbol{h}^T(t_2)\right)\right\}. \tag{5.5}$$

Let $\Delta t = t_2 - t_1$ be the time-lag between two channel snapshots at time instances t_1 and t_2, respectively. Assuming that $\boldsymbol{h}(t)$ is WSS in the slow-time domain, then the radio channels with a vanishing CCF $\gamma_{\boldsymbol{h}}(\Delta t) \approx 0$ correspond to the case when individual taps in CIR are circularly symmetric complex Gaussian random variables. This is usually the case for norrow-band rich scattering channels.

Temporal Auto-Correlation Function (ACF)

The temporal ACF between two channel snapshots at time instances t_1 and t_2 is defined as

$$c_{\boldsymbol{h}}(t_1,t_2) = E\left\{\boldsymbol{h}(t_1)\boldsymbol{h}^*(t_2)\right\} \tag{5.6}$$

where $\boldsymbol{h}^*(t)$ is the complex conjugate of $\boldsymbol{h}(t)$. For $\Delta t > 0$, the absolute value of $c_{\boldsymbol{h}}(\Delta t)$ quantifies the channel coherence time (or distance) and its knowledge plays a key role in the channel tracking algorithms [100]. In general, the realistic channels exhibit asymmetrical Doppler power spectrum density (PSD). Same is true if channels are illuminated with directional antennas. Therefore, the resulting ACF is expected to be complex valued. Let $\mu_{II}(\Delta t)$ and $\mu_{QQ}(\Delta t)$ be the ACFs of the inphase and quadrature components respectively, then real and imaginary parts of $c_{\boldsymbol{h}}(\Delta t)$ are defined as

$$\Re\left\{c_{\boldsymbol{h}}(\Delta t)\right\} = \mu_{II}(\Delta t) + \mu_{QQ}(\Delta t) \tag{5.7}$$

and

$$\Im\{c_{\boldsymbol{h}}(\Delta t)\} = \gamma_{IQ}(\Delta t) - \gamma_{QI}(\Delta t)\cdot \tag{5.8}$$

Complementary Auto-Correlation Function (CACF)

The CACF defines a correlation between $\boldsymbol{h}(t)$ and $\boldsymbol{h}^*(t)$ and it is described as

$$r_{\boldsymbol{h}}(t_1,t_2) = E\left\{\boldsymbol{h}(t_1)\boldsymbol{h}^T(t_2)\right\}\cdot \tag{5.9}$$

Notice that $\boldsymbol{h}(t)$ is second-order stationary if CACF is only a function of time difference Δt. In such a case, the real and imaginary parts of the CACF are defined as

$$\Re\{r_{\boldsymbol{h}}(\Delta t)\} = \mu_{II}(\Delta t) - \mu_{QQ}(\Delta t) \tag{5.10}$$

and

$$\Im\{r_{\boldsymbol{h}}(\Delta t)\} = \gamma_{IQ}(\Delta t) + \gamma_{QI}(\Delta t)\cdot \tag{5.11}$$

In case of Rayleigh-Rice multipath channels, both $r_{\boldsymbol{h}}(\Delta t)$ and $\gamma_{\boldsymbol{h}}(\Delta t)$ are zero $\forall \Delta t$ due to the reason that

$$\mu_{II}(\Delta t) = \mu_{QQ}(\Delta t) \tag{5.12}$$

and

$$\mu_{IQ}(\Delta t) = \mu_{QI}(\Delta t) = 0\cdot \tag{5.13}$$

In the case of sparse radio channels, both conditions shown in Eqs. (5.12) and (5.13) cannot be satisfied [13] and the resulting CACF would then be a non-zero complex valued function. This can be confirmed from the experimental results in Fig. 5.6 which demonstrate that the CCF $\gamma_{IQ}(\Delta t)$ is non-negligible for the whole time-lag $\Delta t > 0$. Such a result is well expected because the directional antennas act as spatial filters for a multipath channel which results in a sparse radio channel. Additionally, the Fig. 5.6 also shows that the CACF values $r_{\boldsymbol{h}}(\Delta t)$ converge towards zero when Rx antenna directivity is reduced—

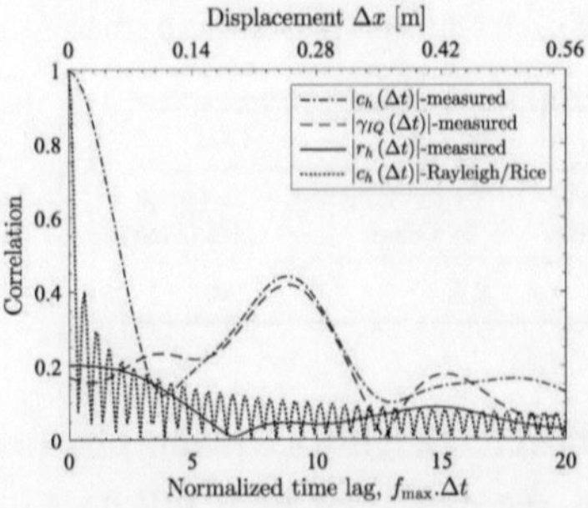

(a). NLOS, Tx 15° HPBW, Rx 15° HPBW

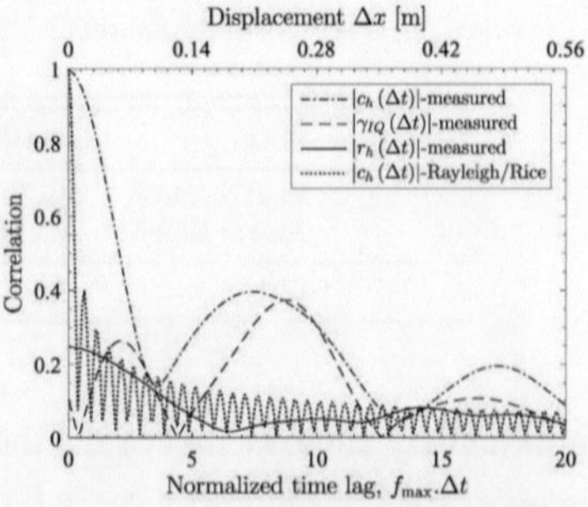

(b). NLOS, Tx 15° HPBW, Rx 30° HPBW

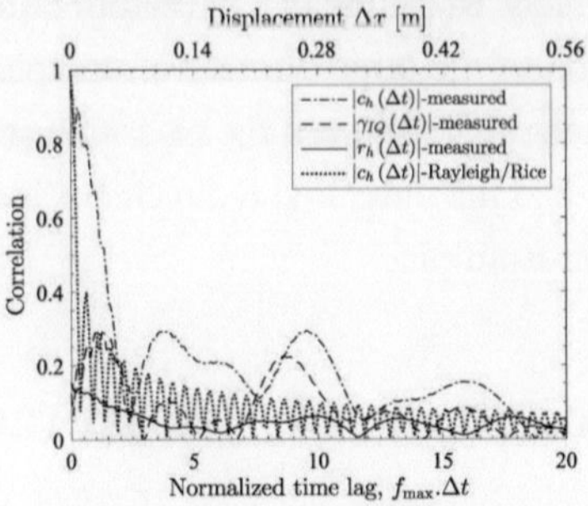

(c). NLOS, Tx 15° HPBW, Rx Omni

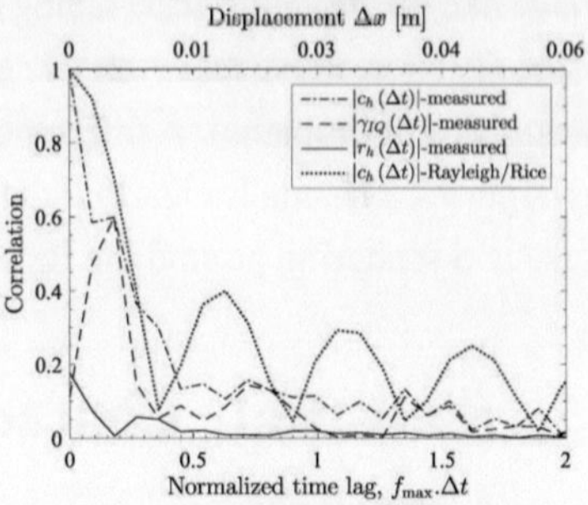

(d). LOS, Tx Omni and Rx Omni

Fig. 5.6. Analysis of the temporal correlation functions for different measurement setups, $f_{\max} = 100$ Hz, VV polarization. Notice that, for Rayleigh-Rice multipath channels $c_{\boldsymbol{h}}(\Delta t) = J_0(2\pi f_{\max}\Delta t)$, where J_0 is a zeroth-order Bessel function of first kind.

explaining increased scattering. Similar results have been demonstrated by a recent experimental work from Hughes et. al. in [105].

From Fig. 5.6, it is also clear that when the antenna directivity (or system beamforming gain) is reduced, the temporal ACFs of the measured channels converge towards the Rayleigh-Rice multipath channel. Considering that a channel is correlated if the absolute value of $c_{\boldsymbol{h}}(\Delta t) \geq 0.5$, then Table 5.3 shows correlation distances (d_{corr}) for different measurement setups. As expected the value of d_{corr} reduces with the low gain antennas as pointed out by Clarke in [45]. In the case of omni-directional antennas at both Tx and the Rx, the results in Fig. 5.6d show that d_{corr} of the measured channel is

Table 5.3. Analysis of correlation distance (d_{corr}) for different measurement setups

Visibility	NLOS	NLOS	NLOS	LOS
	Tx 15° HPBW-Rx 15° HPBW	Tx 15° HPBW-Rx 30° HPBW	Tx 15° HPBW-Rx Omni	Tx Omni-Rx Omni
d_{corr}	12 λ_c	10 λ_c	8 λ_c	1 λ_c

approximately same as that of the theoretical Rayleigh-Rice fading channels. However, for the distances $\Delta x \geq 0.01$ m, the measured ACF still does not behave like the Rayleigh-Rice fading channels. Recalling that the zeroth-order Bessel function is basically an integration of the large number of complex sinusoids (or multipaths). A difference in the ACF between the measured and Rayleigh-Rice channels clearly indicates that the measured radio channels is sparse and scattering around the Rx is non-isotropic.

5.5 Statistical Characterization of Measured Channels

Objective of statistical characterization, in this section, is to study the impact of high antenna directivity and system bandwidth on WSS of the Rx signal. A random process is said to be WSS if its mean, variance and auto-correlation function (ACF) are time– or shift– invariant.

5.5.1 WSS in the Rx Signal Magnitude

By definition, a stationary random process always return towards its mean value and it does not show any specific increasing/decreasing trend. Consequently, the mean value of a stationary process remains constant over the whole observation interval. Intuitively, when small-scale fading dominates over the large-scale fading effects in a channel, then a deterministic increasing/decreasing trend (due to path-loss) in the Rx magnitude may vanish. Consequently, the Rx signal magnitude appears to be a stationary random process

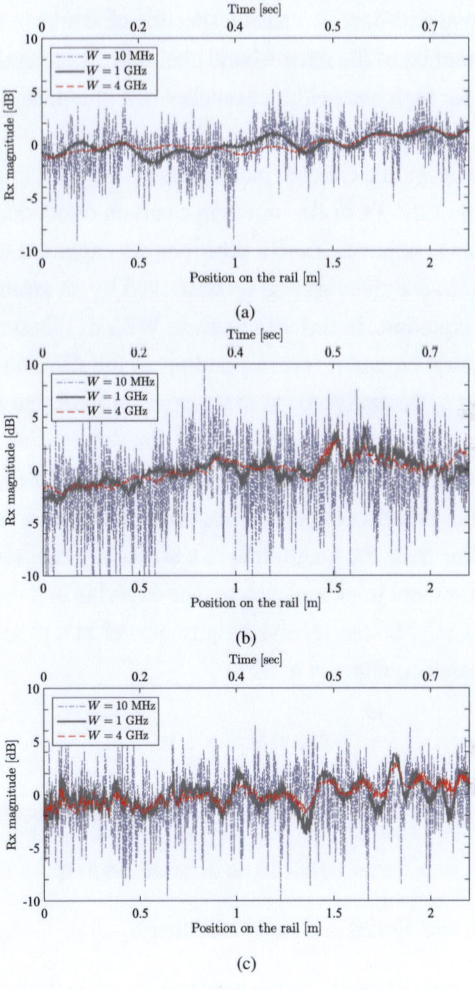

Fig. 5.7. Rx signal magnitude (relative to the 0 dB mean value) analysis over the whole measurement track, VV polarization, speed of the Rx is equal to 10 kilometer per hour, (a) NLOS, Tx 15° HPBW, Rx 15° HPBW. (b) NLOS, Tx 15° HPBW, Rx-Omni. (c)LOS, Tx-Omni, Rx-Omni.

within a fading interval. On the other hand, in the absence of small-scale fading, the Rx signal magnitude is expected to be a non-stationary process due to the dominance of path-loss. A visual inspection of results in Fig. 5.7 demonstrates that, in contrast to the narrowband channels, Rx magnitude shows an increasing trend for high bandwidth channels when illuminated with high gain antennas.

In general, propagation channel parameters evolve with time and displacement when either of the Tx or Rx moves in a certain direction. Consequently, channel snapshots at adjacent Tx–Rx positions are supposed to be correlated. This results in a time series that can be described by an arbitrary difference (or differential) equation. In order to analyze WSS of this time series, roots of the characteristic equation (corresponding to the difference equation of time series) need to be analyzed. If there exists at least one *unit root*[2] then it implies that mean, variance and ACF are time-variant which result in a non-WSS time series. We use the Dickey-Fuller (DF) test [106] with a null hypothesis that the process contains at least one unit root. A rejection of null hypothesis implies that, Rx magnitude is a stationary random process over the whole measurement track and acceptance explains that the time series is non-stationary. Let y_t be the received signal power at a time instant t, then following regression equation is used:

$$\Delta y_t = \Psi y_{t-1} + u_t \tag{5.14}$$

where u_t describes the error term and the following hypothesis are tested

- H_0: $\Psi = 0 \Longrightarrow$ Series needs to be differenced to make it stationary
- H_1: $\Psi < 0 \Longrightarrow$ Series is already stationary.

Accept (A) and reject (R) decisions are based on the critical value (CV) of -3.43 (1 % significance), where the values lower than the CV result in a rejection of null hypothesis and vice versa. As expected, test statistics in Table 5.4 show that the null hypothesis is rejected (marked in BLUE) in cases when system bandwidth and directivity of Tx-Rx antennas are low. Additionally,

[2] Solution set to the characteristic equation contains a root with an absolute value equal to 1.

Table 5.4. Results obtained from Dickey-Fuller test

Visibility	Setup	Absolute bandwidth (GHz)								
		0.01	0.05	0.1	0.3	0.5	0.7	1	2	4
NLOS	Rx 15° HPBW, Tx 15° HPBW	-23.76 (R)	-12.05 (R)	-7.1 (R)	-3.2 (A)	-2.7 (A)	-2.4 (A)	-2.2 (A)	-2.1 (A)	-2.08 (A)
NLOS	Rx 30° HPBW, Tx 15° HPBW	-24.84 (R)	-10.83 (R)	-8.7 (R)	-3.1 (A)	-2.6 (A)	-2.3 (A)	-2.04 (A)	-1.7 (A)	-2.01 (A)
NLOS	Rx Omni, Tx 15° HPBW	-26.7 (R)	-21.4 (R)	-17 (R)	-17.7 (R)	-14.3 (R)	-8 (R)	-4.4 (R)	-3.3 (A)	-3.3 (A)
LOS	Rx Omni, Tx Omni	-28.6 (R)	-21 (R)	-24.5 (R)	-15.9 (R)	-12 (R)	-9.5 (R)	-8.5 (R)	-8.4 (R)	-7.6 (R)

these results show that for a certain Tx-Rx antenna directivity null hypothesis is rejected at a particular increase in system bandwidth. This implies that unit root exists in the data series and Rx signal magnitude is not a stationary random process for system configuration where bandwidth and antenna directivity is high.

In Table 5.5, the p and R^2 values of the regression process are shown to analyse if there exists a trend and its statistical significance. A $p \leq 0.05$ indicate that an increasing/decreasing trend in Rx signal magnitude is statistically significant. On the other hand, the $R-$square values range from $0 \leq R^2 \leq 1$ where the values closer to 1 indicate goodness of the fit and vice versa. Results in Table 5.5 show that $p-$value is zero in each case which demonstrate that there exists a statistically significant trend in the Rx signal magnitude. However, the R^2 values are quite low. This explains that although there exist a statistical trend but the regression does not satisfy the goodness-of-fit criteria $R^2 \geq 0.54$ particularly in cases where system bandwidth and antenna directivity is low. This also explains why DF test rejects the null hypothesis. A statistical significant trend as explained by $p = 0$ is due to the reason that measurement track is quite long i.e. 2.2 meters and Rx- signal magnitude may increase/decrease significantly due to variations in the link distance.

Table 5.5. p and R^2 values of the regression tests (p,R^2)

Visibility	Setup	Absolute bandwidth (GHz)								
		0.01	0.05	0.1	0.3	0.5	0.7	1	2	4
NLOS	Rx 15° HPBW, Tx 15° HPBW	(0,0.08)	(0,0.3)	(0,0.4)	(0,0.45)	(0,0.45)	(0,0.47)	(0,0.5)	(0,0.6)	(0,0.64)
NLOS	Rx 30° HPBW, Tx 15° HPBW	(0,0.01)	(0,0.35)	(0,0.4)	(0,0.5)	(0,0.5)	(0,0.5)	(0,0.54)	(0,0.67)	(0,0.7)
NLOS	Rx Omni, Tx 15° HPBW	(0,0.05)	(0,0.18)	(0,0.2)	(0,0.2)	(0,0.24)	(0,0.4)	(0,0.43)	(0,0.5)	(0,0.57)
LOS	Rx Omni, Tx Omni	(0,0.02)	(0,0.09)	(0,0.1)	(0,0.1)	(0,0.1)	(0,0.1)	(0,0.14)	(0,0.4)	(0,0.54)

5.5.2 Variance of the Rx Signal Magnitude

In general, a small-scale fading area is defined such that channel statistics does not vary considerably and the Rx signal remains a stationary random process. An exact length of the stationary interval is hard to define [107], because radio channel statistics are strongly coupled with the properties of system under consideration, as shown in this paper. Length of the stationarity (i.e., $30\lambda_c$) interval, in this work, is largely based on the visual inspection which may not be very accurate for each measurement setup. A smaller length $(< 30\lambda_c)$ may be more appropriate, but $30\lambda_c$ is chosen because authors wished to have enough measured channel snapshots as well, for better statistical accuracy. For this purpose, an overlapping sliding window of length $30\lambda_c$ is moved over the whole 2.2m long measurement track. This results in an extraction of 822 small-scale fading areas each of length $30\lambda_c$. For each of these intervals, irrespective of any W, mean value of $E_{\mathrm{Rx}}(t)$ is maintained as $\rho = 0$ dB due to the normalization. Thus allowing only the standard deviation (σ_0) of $E_{\mathrm{Rx}}(t)$ to change along different fading intervals and for different measurement setups. Intuitively, a high similarity of σ_0 values over the measurement track indicates that, σ_0 is time-invariant (a requirement of WSS) and vice versa. Two interesting observations are evident from results in Fig. 5.8,

- Fading of the Rx signal magnitude vanishes because the average spread $(\bar{\sigma}_0)$ reduces with W up to a bandwidth $W_0 = 2.5$ GHz, known as stability

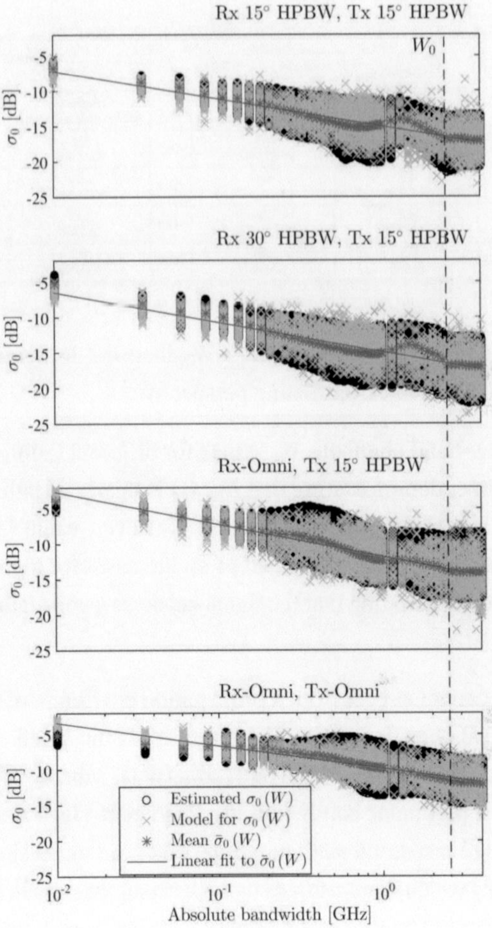

Fig. 5.8. Measured and modeled values of $\sigma_0(W)$ while maintaining $\rho = 0$ dB mean value due to the channel normalization.

Table 5.6. Proposed models for the mean $\bar{\sigma}_0(W)$ and standard deviation $\mathscr{X}_{\sigma_0}(W)$ of σ_0 and their parameter values

Visibility	Measurement setup	Model and parameter values $\bar{\sigma}_0(W) = \begin{cases} c_0 - m\log_{10}W, & \text{if } W \leq W_0 \\ c_1 & \text{otherwise} \end{cases}$			Model and parameter values $\mathscr{X}_{\sigma_0}(W) = c_0 + m\log_{10}W$	
		Slope m (95% confidence bounds)	Intercept c_0 (95% confidence bounds)	c_1	Slope m (95% confidence bounds)	Intercept c_0 (95% confidence bounds)
NLOS	Rx 15° HPBW, Tx 15° HPBW	3.9 (4.2,3.7)	-15 (-15.1,-14.9)	-16.7	0.5 (0.36,0.6)	1.7 (1.66,1.8)
NLOS	Rx 30° HPBW, Tx 15° HPBW	4.2 (4.5, 3.8)	-14.9 (-15.1,-14.7)	-16.7	0.6 (0.54,0.7)	1.86 (1.82,1.9)
NLOS	Rx Omni, Tx 15° HPBW	4.6 (4.8, 4.4)	-11.8 (-11.9,-11.7)	-13.8	0.63 (0.4,0.8)	2 (1.9,2.1)
LOS	Rx Omni, Tx Omni	3.2 (3.5,3)	-9.9 (-10,-9.7)	-11.2	0.5 (0.4,0.6)	1.28 (1.2,1.33)

bandwidth [42]. For bandwidth $W > W_0$, there are no significant changes in the radio channel from fading perspectives.

- For narrow-band channels, σ_0 values for different fading intervals are almost same, demonstrating that $E_{\text{Rx}}(t)$ is nearly a stationary process over the whole measurement track. However, same is not true for wide-band channels as the spread of σ_0 increases for higher bandwidth channels demonstrating that Rx signal becomes a non-stationary random process.

Intuitively, one cannot expect from a propagation environment to favour one stationary interval over the other. Therefore, σ_0 is modelled as a normally distributed random process $\mathcal{N}\left(\bar{\sigma}_0(W), \mathscr{X}_{\sigma_0}(W)\right)$, where, $\mathscr{X}_{\sigma_0}(W)$ is the spread of σ_0 at a particular bandwidth W. Results in Fig. 5.8 show that, σ_0 obtained from 822 measured stationary intervals is an increasing function of W around $\bar{\sigma}_0(W)$ which demonstrates that Rx signal magnitude becomes non-WSS. This is indicated by the positive slopes m in the slope-intercept model of $\mathscr{X}_{\sigma_0}(W)$ in Table 5.6. One may note that, an increase in $\mathscr{X}_{\sigma_0}$ is not monotonic with W. However, an approximation of $\mathscr{X}_{\sigma_0}(W)$ with a slope-intercept form in Table 5.6 is done to maintain simplicity and intuitiveness. Fig. 5.8 shows that, the slope-constant model with parameter values in Table 5.6 provides a very good approximation of $\bar{\sigma}_0(W)$ and same is the case for slope-intercept model derived for variance of $\sigma_0(W)$.

5.5.3 WSS in the Complex CIR Over the Slow-Time

Now we intend to analyze if the channel temporal ACF $c_h(\Delta t, \tau) = E[h(t+\Delta t, \tau) h^*(t, \tau)]$ is shift- or time-invariant—a requirement for CIR to be WSS in the slow time. Let $\mu_{II}(\Delta t, \tau)$ and $\mu_{QQ}(\Delta t, \tau)$ are ACFs of inphase (I) and quadrature (Q) components and $\gamma_{IQ}(\Delta t, \tau)$ is their temporal cross-correlation, then real ($\Re$) and imaginary ($\Im$) parts of $c_h(\Delta t, \tau)$ are defined as

$$\Re\{c_h(\Delta t, \tau)\} = \mu_{II}(\Delta t, \tau) + \mu_{QQ}(\Delta t, \tau) \quad (5.15)$$

and

$$\Im\{c_h(\Delta t, \tau)\} = \gamma_{IQ}(\Delta t, \tau) - \gamma_{QI}(\Delta t, \tau) \cdot \quad (5.16)$$

where,

$$\mu_{II}(\Delta t, \tau) = E[h_I(t+\Delta t, \tau) h_I^*(t, \tau)] \quad (5.17)$$

$$\mu_{QQ}(\Delta t, \tau) = E\left[h_Q(t+\Delta t, \tau) h_Q^*(t, \tau)\right] \quad (5.18)$$

$$\gamma_{IQ}(\Delta t, \tau) = E\left[h_I(t+\Delta t, \tau) h_Q^*(t, \tau)\right] \quad (5.19)$$

$$\gamma_{QI}(\Delta t, \tau) = E[h_Q(t+\Delta t, \tau) h_I^*(t, \tau)] \cdot \quad (5.20)$$

A real valued $c_h(\Delta t, \tau)$ is possible when $\gamma_{IQ}(\Delta t, \tau) = \gamma_{QI}(\Delta t, \tau) = 0, \forall \Delta t \geq 0$, which leads to a fact that complex Rx signal is a circularly symmetric complex Gaussian random variable and the channel corresponds to a rich scattering environment [101]. Using the same measurement data, validity of the narrowband rich scattering assumption has been analyzed in our earlier work [31]. Results in [31] demonstrated that rich scattering assumption does not remain valid due to high gain antennas and higher bandwidth of mmWave systems. Consequently, Rayleigh–Rice fading channels for the cluster fading envelope are not realistic.

In general, for non-Gaussian signals, second-order statistics of a time domain signal are not fully described by $c_h(\Delta t, \tau)$ [101, 102]. Therefore, another function, denoted either as relation function $r_h(\Delta t, \tau) =$

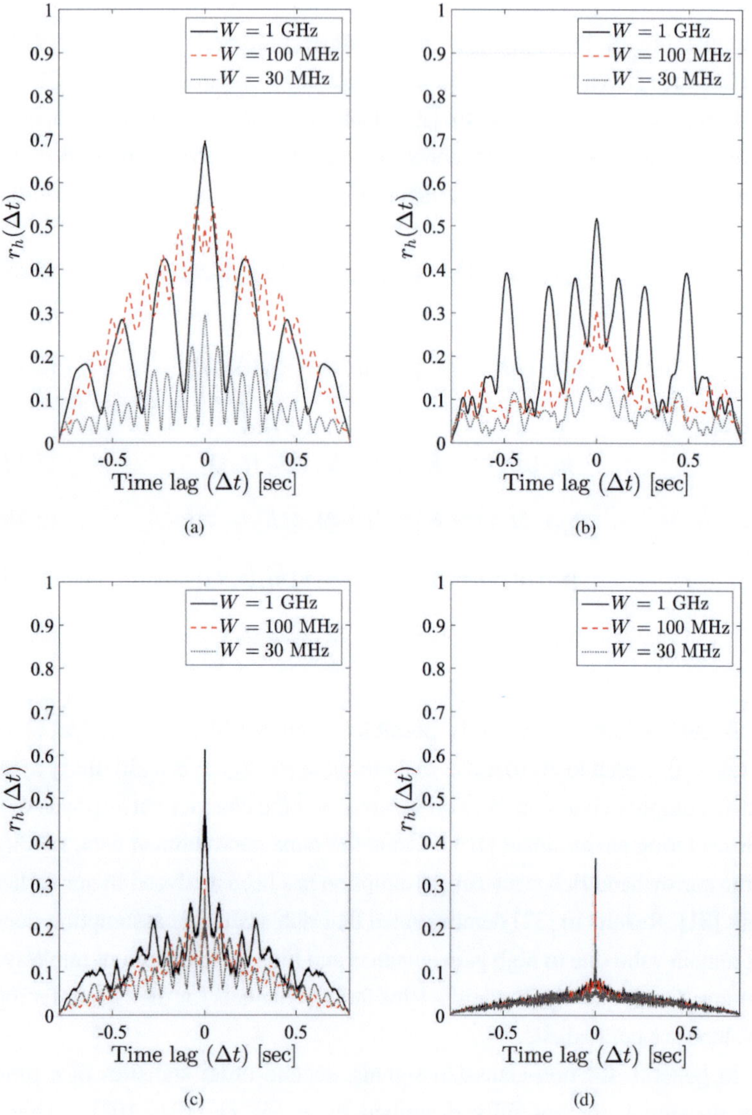

Fig. 5.9. Analysis of CACF (averaged over multipath delays) for WSS testing. Speed of the Rx is equal to 10 kilometer per hour, (a) NLOS, Tx 15° HPBW, Rx 15° HPBW. (b) NLOS, Tx 15° HPBW, Rx 30° HPBW. (c) NLOS, Tx 15° HPBW, Rx-Omni directional. (d) LOS, Tx-Omni, Rx-Omni directional.

$E[h(t+\Delta t,\tau)h(t,\tau)]$ [101] or complementary ACF (CACF) [102] is used which measures a correlation between $h(\Delta t,\tau)$ and $h^*(\Delta t,\tau)$. From hereinafter, in the subsequent discussions, $r_h(\Delta t,\tau)$ is referred to as CACF. The real and imaginary parts of the CACF are defined as

$$\Re\{r_h(\Delta t,\tau)\} = \mu_{II}(\Delta t,\tau) - \mu_{QQ}(\Delta t,\tau) \tag{5.21}$$

and

$$\Im\{r_h(\Delta t,\tau)\} = \gamma_{IQ}(\Delta t,\tau) + \gamma_{QI}(\Delta t,\tau)\cdot \tag{5.22}$$

A CACF provides following insights into the channel properties:

- $r_h(\Delta t = 0,\tau) = 0$ implies that the channel $h(t,\tau)$ is a circularly symmetric random variable—a requirement for Rayleigh–Rice fading signals.
- Vanishing of CACF $r_h(\Delta t,\tau) \approx 0$, $\forall \Delta t$ implies that $c_h(\Delta t,\tau)$ is time-invariant—a requirement for WSS signal [108, 109]. Therefore, following two conditions must be satisfied,

$$\mu_{II}(\Delta t,\tau) = \mu_{QQ}(\Delta t,\tau) \tag{5.23}$$

and

$$\gamma_{IQ}(\Delta t,\tau) = -\gamma_{QI}(\Delta t,\tau)\cdot \tag{5.24}$$

These conditions imply that, for a CIR to be WSS in slow time, the I and Q components must be balanced in the sense that their ACFs are the same and their cross-correlation function $\gamma_{IQ}(\Delta t,\tau)$ is an odd function. For each measurement setup, results in Fig. 5.9 show CACFs (averaged over path delays) for different bandwidth channels. It is interesting to note that, when high gain antennas are used, $r_h(\Delta t)$ does not vanish to zero even for narrowband channels (e.g. 30 MHz and 100 MHz). In contrast, an increasing trend in values of $r_h(\Delta t)$ is quite evident with an increase in bandwidth and antenna gain. However, when omni-directional antennas are used, the measured CIR demonstrates properties quite closer to a WSS signal. As expected, an increase in system bandwidth also have a profound impact on the complex CIR which becomes non-WSS in high bandwidth channels.

5.5.4 WSS in Frequency Domain

For a time varying CTF $H_W(t,f)$, this section aims to analyze WSS in the frequency domain; uncorrelated scattering in the delay domain. This means that, at a particular time instant t, the channel frequency correlation function (FCF)

$$c_H(t,\Delta f) = E\left[H(t,f+\Delta f)H^*(t,f)\right] \tag{5.25}$$

depends only on the frequency lag $\Delta f = f_2 - f_1$. This means that, $c_H(t,\Delta f)$ does not depend upon the absolute frequency f. Similar like complex Gaussian CIR, $H(t,\Delta f)$ is a complex Gaussian random process in frequency (hence WSS), if the second-order statistics are fully described by $c_H(t,\Delta f)$ and complementary auto-correlation function

$$r_H(t,\Delta f) = E\left[H(t,f+\Delta f)H(t,f)\right] \tag{5.26}$$

is zero for all $\Delta f \geq 0$. All results in Figs. 5.10a to 5.10d show that, the time averaged $r_H(\Delta f) \approx 0$ for the whole frequency lag Δf. This implies that the function $c_H(t,\Delta f)$ in (5.25) is independent of absolute frequency and measured channels are WSS in the frequency domain.

Assuming that, the frequency domain samples in the CTF are decorrelated when $c_H(t,\Delta f) \leq 0.5$, then the coherence bandwidth B_c evaluated from the measured channels is shown in Fig. 5.10. As expected from the delay spread statistics in Fig. 5.4, the coherence bandwidth is quite large i.e., $B_c = 0.2$ GHz in the case when highly directive 15° HPBW Tx-Rx antennas are employed. However, coherence bandwidth reduces when less directive antennas are used at Rx. In the LOS case, when omni-directional antennas at both Tx-Rx are used to illuminate the channel, coherence bandwidth is lowest, i.e., $B_c = 0.12$ GHz.

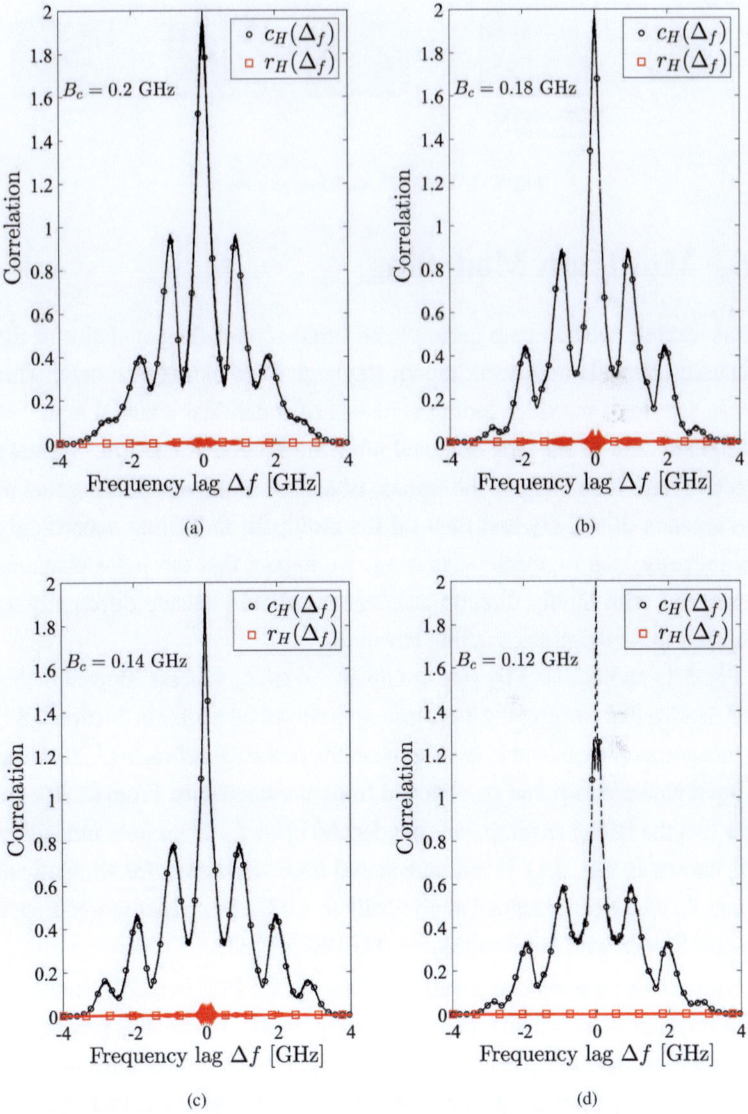

Fig. 5.10. Correlation analysis of CTF (averaged over each time snapshot) for the testing of WSS in frequency domain. Speed of the Tx is equal to 10 kilometer per hour, (a) NLOS, Tx 15° HPBW, Tx 15° HPBW. (b) NLOS, Tx 15° HPBW, Tx 30° HPBW. (c) NLOS, Tx 15° HPBW, Tx-Omni directional. (d) LOS, Tx-Omni, Tx-Omni directional.

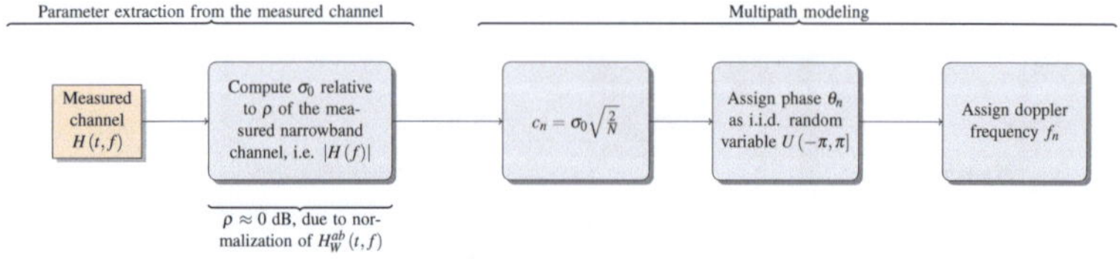

Fig. 5.11. Multipath modeling process

5.6 Multipath Modelling

In this section, we compare narrowband small-scale fading statistics of the measured channels with well known Rayleigh-Rice fading channels. This means, we don't consider fading statistics of individual channel taps, but a weighted sum of all taps obtained after the Fourier transform. Primary objective here is to analyse the impact of spatial multipath filtering due to high antenna directivity and then do the multipath modelling accordingly. Theoretically, due to sparse scattering, we expect that the radio channels illuminated with highly directional antennas should behave differently as compared to Rayleigh-Rice fading envelops.

Fig. 5.11 shows a step-by-step multipath modelling process adopted in this work. Notice that, we employ SOC principle described earlier in Section 2.5.1. For this purpose, we intend to fit the probability density function in (2.20) using the input parameters ρ and σ_0 obtained from measurements. From (2.20), we know that the fading envelop does not depend upon f_n. Therefore, modelling of f_n shown in Fig. 5.11 is not considered here. However, for simulations we use f_n, which are assigned analytically to MPCs using Method-of-Exact-Doppler-Spread with Set-Partitioning (MEDS-SP) [33].

Notice that for particular ρ and $c_n = \sigma_0\sqrt{\frac{2}{N}}$, the PDF in (2.20) converges to the Rayleigh/Rician fading distribution as $N \to 10$. We use this property to understand the scattering behaviour (rich or sparse) and model the amplitude gains c_n and the number of scattered MPCs accordingly. For this purpose, we check the value of N for which the Euclidean distance between the PDF in (2.20) is minimum to measured channel PDF envelop denoted as $P_\xi^{\mathrm{m}}(z)$. This requires extensive evaluations of (2.20) for $N \geq 1$ and comparison with $P_\xi^{\mathrm{m}}(z)$. Fig. 5.12a shows $P_\xi^{\mathrm{m}}(z)$ of two of the measured channel setups and results of the SOC process, both are in excellent agreement with each other for $N = 4$ and $N = 6$.

The quality of approximation between $P_\xi^{\mathrm{m}}(z)$ and the theoretical PDF of SOC model in (2.20) can be described by the mean square error (MSE). It is defined as

$$\mathrm{MSE} = \frac{1}{m_z} \sum_{z=1}^{m_z} \left(\underbrace{P_\xi(z)}_{\text{SOC theory}} - \underbrace{P_\xi^{\mathrm{m}}(z)}_{\text{measurement}} \right)^2 \tag{5.27}$$

where, $m_z = 120$ is the total number of samples of measured Rx signal magnitude corresponding to the number measurement positions in a particular stationary interval. Results in Fig. 5.12b show that MSE reduces with an increase in the number of cisoids. For smaller values of N, PDF in (2.20) may result in an unknown distribution which could be a better fit to $P_\xi^{\mathrm{m}}(z)$ rather than Rayleigh/Rician fit. This has been demonstrated in two Tx-Rx setups $(15^\circ - 15^\circ$ and $15^\circ - 30^\circ)$ in Fig. 5.12b, where $p_\xi(z)$ fits better to $P_\xi^{\mathrm{m}}(z)$ for lower values of N. For higher values of N, MSE increases as $p_\xi(z)$ converges to Rayleigh/Rician fit of $P_\xi^{\mathrm{m}}(z)$. From Fig. 5.12b, it is easier to follow that channels with beamforming at the Rx can be effectively modelled with $N = 2 - 3$ scatterers/sinusoids. This implies that, fading envelope derived from non- Gaussian complex random variable assumption fit better to the statistical properties of $H(f)$. On the other hand, channels with omni-directional Rx antenna require increased number of scattered components i.e. $N = 4 \rightarrow 6$ in the considered scenario.

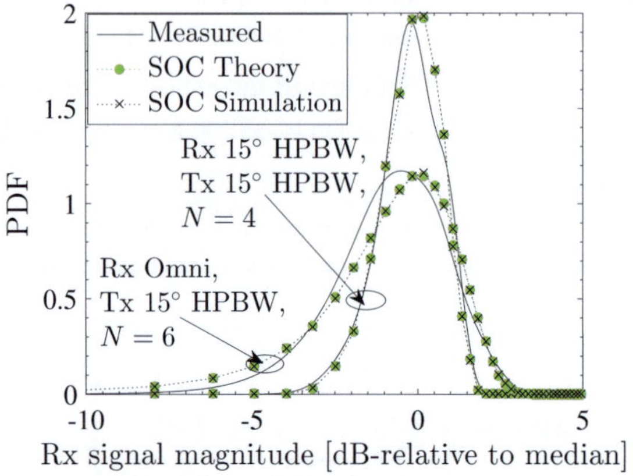

(a)

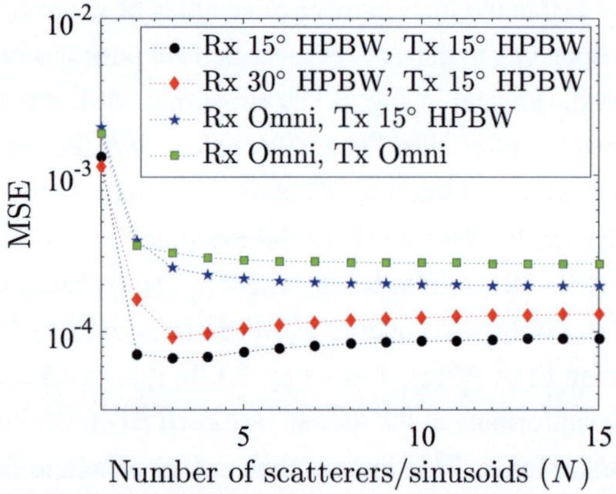

(b)

Fig. 5.12. (a) Comparison of the PDFs of measured and modelled channels and (b) MSE; of the measured and modelled channels using SOC principle at $W = 10$ MHz.

5.7 Conclusion

This chapter adds a further value to the results shown in Chapter 4, where, in addition to system bandwidth the impact of antenna directivity has also be explained. Sparse multipath illumination with highly directive antennas impacts the radio channel properties significantly. For example, fading depth reduces which implicitly explains that fading of taps vanishes. Additionally, the complex CIR vector of a cluster does not remain WSS in the slow time domain. This leads to a conclusion that complex CIR vector for a cluster cannot be modelled as a Rayleigh-Rice multipath channel.

Chapter 6

Summary and Future Work

In this thesis, an effort has been made to identify the research areas where channel modelling for milli-meter (mmWave) radio channels may differ from well known narrowband channel models known from sub-6 GHz systems. Our study of the state-of-the-art (SOTA) channel models for mmWave systems reveals that these channel models still rely on narrowband assumptions inherited from legacy sub-6 GHz channel models [10–12]. Notice that, primary motivation behind these assumptions comes from low antenna directivity and small bandwidth of the sub-6 GHz systems. In contrast, mmWave systems are different in the sense that antenna directivity is supposed to be high such that higher path losses at these frequencies can be mitigated. Additionally, to achieve high data rates, a mmWave system is supposed to operate with a high bandwidth which is available at these frequencies. Communication links are supposed to be established through individual multipath components in the channel due to spatial multipath filtering through high gain antennas. Consequently, each individual resolvable MPC and its associated parameters such as its amplitude gain, phase, polarization and Doppler frequency modelling become a subject of high importance.

During the course of this work, together with colleagues from mmMAGIC project [58], significant efforts have been made to understand the frequency dependency of propagation. Apart from the results presented in this work, extensive measurement campaigns have been carried out in wide range of scenarios including outdoors, indoors and outdoor-to-indoors. In this work, the analysis of multipath dispersion statistics align fully with conclusions drawn in mmMAGIC project [58] that frequency has a limited impact on the channel model definition. Main difference, in general, comes from differences in the scenario due to different propagation environment. Otherwise, multipath dispersion statistics are more are less same if the environment and assumptions on the antenna directivity, system bandwidth and its dynamic threshold are same.

In general, it is well understood that delay and angular spreads reduce with an increase in antenna directivity. Fade depth, in general, also reduces with an increase in system bandwidth [42, 43]. During this work, we have learned that fade depth reduces with an increase in antenna directivity as well [26]. Fading behaviour of each multipath cluster from a different reflection surface differs considerably, demonstrating that mmWave frequencies are more sensitive to reflection surfaces [30, 24]. We have proposed a simple but accurate model that translates fade depth reduction for different reflection surfaces while taking into account the antenna directivity and system bandwidth [26, 30, 24].

Theoretically, variance of the resolvable channel taps reduces with an increase in the system bandwidth and the tap becomes temporally sparse. In contrast to the SOTA channel models for mmWave systems, such a tap cannot be modelled as a complex Gaussian random variable [13]. When such a tap is illuminated with a highly directive antenna, then a frequency bins of the complex channel transfer function cannot be modelled as a complex Gaussian random variable. We have demonstrated and modelled this effect with SOC principle [27] using tap mean and variance values obtained from channel measurements (see Section 5.6). From channel modelling point of view, illuminating a non-Gaussian tap implies that capacity or mutual information of the such channels would be quite lower particularly in scenarios with a high signal-to-noise ratio.

Resolvable multipaths in SOTA channel models for mmWave systems are modelled such that their phase is a uniformly distributed variable over

$U[-\pi, \pi]$. A reduction in the variance of such multipaths, as mentioned above, implicitly leads towards an increase in K-factor value. In such a case, from the theory of SOC principle [27], we have learned that the phase of such a channel tap converges to be a non-uniform deterministic process. Modelling the phase of such channel taps can be done with SOC principle using tap mean and variance values obtained from measurements as shown in this work.

In contrast to SOTA channel models, our work highlights that given the same angle-of-incidence and frequency, average XPR values for each multipath cluster differ significantly [28]. Additionally, randomness in the XPR vanishes exponentially with an increase in bandwidth. For different reflection surfaces, we have provided average XPR values and their variances as model parameters for system bandwidth ranging from 20 MHz to 4 GHz.

After knowing that resolvable multipaths cannot be modelled as complex Gaussian random variables, we focused on the analysis of whole complex CIR vector. Due to the Rayleigh-Rice fading assumptions on each multipath, complex CIR vectors in SOTA channel models for mmWave becomes a complex Gaussian random vector. In our work [30, 31], it has been shown that complex valued CIR vector cannot be modelled as Gaussian random vector particularly in directional radio channels. Here again directivity of the antenna plays a significant role in such a way that propagation channels illuminated with highly directional antennas demonstrate significantly different correlation properties compared to that of omni-directional antennas. Multipath channels illuminated with omni-directional antennas demonstrate similar correlation properties as that of Rayleigh-Rice multipath channels. It has been shown that there exist a correlation (although small) between the resolvable channel taps even at very high bandwidths of 4 GHz (see Section 4.5).

For a stochastic process to be wide-sense-stationary (WSS) implies that its mean and auto-correlation function should be time invariant [33]. Due to the narrow-band Rayleigh fading assumption on the resolvable channels taps, SOTA channel models for mmWave are WSS in both slow-time and frequency domains. A reduction in the variance of a resolvable channel tap with an increase in antenna directivity and bandwidth means that its mean value may not constant when either of Tx or Rx moves in a certain direction. Additionally, the ACFs of inphase and quadrature components may not be equal demonstrating time dependency of the tap ACF. If individual resolvable taps are not WSS, similar impact should appear on the complex CIR vector. Our results in [34] demonstrate that CIR vector of radio channel illuminated with highly directional antennas and high system bandwidths does not remain WSS in the slow-time domain. However, in the frequency domain, they can still modelled as WSS random processes because the correlation between channel taps is quite small. Here one may note that non-Gaussian complex random variables can also be WSS in nature as pointed out in [101]. This will result in over estimation of the channel capacity.

From the results and discussions above, we conclude that the channel modelling methodology in 3GPP-TR 38901 [14] (inherited from [10–12]) and other SOTA channel models does not meet theoretical expectations imposed by high antenna directivity and bandwidth of mmWave systems. Therefore, for accurate performance prediction of mmWave systems, channel modelling methodology in SOTA channel models need to be revised. These revisions and concept validations must be supported by extensive measurements and data evaluations.

As a future work, we propose to investigate the following research directions:

(i) Experiments with Different Reflection Surfaces. Notice that results presented in this work are based on a limited data set. A promising direction for future research would be to carry out similar investigations and statistical analysis for a wide range of reflection surfaces. This may include smooth and cluttered reflecting objects so that the best and worst case fading behaviour can be analysed and modelled. We expect such a cluttered environment in industrial internet-of-things (IOT) channels. Therefore, these investigations may provide useful insights for tunning channel models derived for conventional cellular and industrial IOT applications.

(ii) Experiments for V2V Communication Links. V2V radio channels are interesting in the sense that both ends of a link are supposed to move. During this work, we have learned that if LOS path is blocked by an overtaking vehicle, a communication link between two vehicles can still be established through other reflection objects from the propagation environment. Fading analysis of such radio links would be an interesting topic to investigate particularly when both ends of thc link arc moving. It is quite plausible that strong flashing waves from other moving cars contribute to fading behaviour of these links by suddenly appearing and disappearing. Impact of such flashing rays in the fading statistics would be an interesting subject to investigate.

(iii) Link Level Performance Analysis. Assuming that the mmWave communication link is established through a single resolvable multipath, it would be interesting to compare performance of SOTA mmWave system using different multipath modelling approaches. For example, one can analyse the performance differences when the resolvable tap is modelled as zero-mean complex Gaussian random variable (SOTA way) to the case when it is modelled as non-zero mean complex non-Gaussian random variable (proposed way). Modelling the tap as complex non-Gaussian means that the channel do not exhibit two degrees of freedom compared to the case when a link is modelled as complex Gaussian random variable. On the other hand, multipath channel taps having a non-zero mean result in an increased channel capacity even if they are sparse as shown in [110]. Therefore, an analysis of exact performance trade-off would be an interesting research topic to investigate.

(iv) Improvements in the Standardized 3GPP-TR 38.901 [14] Channel Model. From author's point of view, the number of intra-cluster rays described inEq. (2.3) is the most important parameter that needs to be precisely fixed because of its high sensitivity to system bandwidth and antenna directivity. This will become even more sensitive for future systems with pencil sharp beams. Additionally, the K-factor of sparsely illuminated channels is also a very important aspect that needs to be studied because it highly influences the phase distribution of such a multipath channel as demonstrated in Fig. 1.4. In order to maintain simplicity, standardization activities can consider the introduction of cluster specific lookup tables where the number of intra-cluster multipaths and K-factor are described as a function of antenna directivity and system bandwidth. Simulations for a particular system may randomly pick the clusters with specific properties defined by its bandwidth and antenna directivity and generate a channel accordingly.

References

[1] A. Maltsev, R. Maslennikov, A. Sevastyanov, A. Khoryaev, and A. Lomayev, "Experimental investigations of 60 GHz WLAN systems in office environment," *IEEE Journal on Selected Areas in Communications*, vol. 27, no. 8, pp. 1488–1499, October 2009.

[2] C. Gustafson and F. Tufvesson, "Characterization of 60 ghz shadowing by human bodies and simple phantoms," in *6th European Conference on Antennas and Propagation (EUCAP)*, March 2012, pp. 473–477.

[3] A. Maltsev, "Channel modeling and characterization-MiWEBA," *Deliverable 5.1 EU Contract No. FP7-ICT-608637*, 2014.

[4] H. Steyskal, "Digital beamforming antennas: An introduction," *Microwave J.*, vol. 30, no. 1, pp. 107–124, 1987.

[5] J. D. Fredrick, Yuanxun Wang, and T. Itoh, "Smart antennas based on spatial multiplexing of local elements (smile) for mutual coupling reduction," *IEEE Transactions on Antennas and Propagation*, vol. 52, no. 1, pp. 106–114, Jan 2004.

[6] G. Krishnamurthy and K. G. Gard, "Time division multiplexing front-ends for multiantenna integrated wireless receivers," *IEEE Transactions on Circuits and Systems I: Regular Papers*, vol. 57, no. 6, pp. 1231–1243, June 2010.

[7] T. Nishio, Hsiao-Ping Tsai, Yuanxun Wang, and T. Itoh, "A high-speed adaptive antenna array with simultaneous multibeam-forming capability," *IEEE Transactions on Microwave Theory and Techniques*, vol. 51, no. 12, pp. 2483–2494, Dec 2003.

[8] E. Dahlman, S. Parkvall, and J. Skold, *5G NR: The next generation wireless access technology*. Academic Press, 2018.

[9] *Evolved Universal Terrestrial Radio Access (E-UTRA) physical channels and modulation*, 3rd Generation Partnership Project Std. Tech. Spec. 36.211 V10.2.0, 2011.

[10] 3GPP, "Spatial channel model for multiple input multiple output (MIMO) simulations," 3rd Generation Partnership Project (3GPP), Tech. Rep. 3GPP TR 25.996 version 12.0.0 Release 12, 2014.

[11] P. Kyosti, "WINNER II channel models," *IST, Tech. Rep. IST-4-027756 WINNER II D1.1.2 V1. 2*, 2007.

[12] L. Liu, C. Oestges, J. Poutanen, K. Haneda, P. Vainikainen, F. Quitin, F. Tufvesson, and P. D. Doncker, "The cost 2100 MIMO channel model," *IEEE Wireless Communications*, vol. 19, no. 6, pp. 92–99, December 2012.

[13] M. Pätzold and G. Rafiq, "Sparse multipath channels: Modelling, analysis, and simulation," in *IEEE 24th Annual International Symposium on Personal, Indoor, and Mobile Radio Communications (PIMRC)*, Sept 2013, pp. 30–35.

[14] *Study on channel model for frequencies from 0.5 to 100 GHz*, 3rd Generation Partnership Project Std. 3GPP TR 38.901 V1.0.1, Rev. Release 14, Mar. 2017.

[15] A. Maltsev et al., *Channel Models for 60 GHz WLAN Systems*, IEEE Std. Document 802.11-09/0334r8, 2010.

[16] ——, *Channel Models for IEEE 802.11ay*, IEEE Std. 802.11-15/1150r9, May 2016.

[17] R. J. Weiler, M. Peter, W. Keusgen, A. Maltsev, I. Karls, A. Pudeyev, I. Bolotin, I. Siaud, and A.-M. Ulmer-Moll, "Quasi-deterministic millimeter-wave channel models in MiWEBA," *EURASIP Journal on Wireless Communications and Networking*, vol. 2016, no. 1, p. 84, 2016. [Online]. Available: http://dx.doi.org/10.1186/s13638-016-0568-6

[18] C. Gustafson, K. Haneda, S. Wyne, and F. Tufvesson, "On mm-wave multipath clustering and channel modeling," *IEEE Transactions on Antennas and Propagation*, vol. 62, no. 3, pp. 1445–1455, March 2014.

[19] V. Nurmela, A. Karttunen, A. Roivainen, L. Raschkowski, V. Hovinen, J. Y. EB, N. Omaki, K. Kusume, A. Hekkala, R. Weiler *et al.*, "Deliverable D1. 4 METIS channel models," in *Proc. Mobile Wireless Commun. Enablers Inf. Soc.(METIS)*, 2015, p. 1.

[20] M. K. Samimi and T. S. Rappaport, "3-D millimeter-wave statistical channel model for 5G wireless system design," *IEEE Transactions on Microwave Theory and Techniques*, vol. 64, no. 7, pp. 2207–2225, July 2016.

[21] M. Boban, D. Dupleich, N. Iqbal, J. Luo, C. Schneider, R. Müller, Z. Yu, D. Steer, T. Jämsä, J. Li, and R. S. Thomä, "Multi-band vehicle-to-vehicle channel characterization in the presence of vehicle blockage," *IEEE Access*, vol. 7, pp. 9724–9735, 2019.

[22] S. Nguyen, J. Medbo, M. Peter, A. Karttunen, K. Haneda, A. Bamba, R. D'Errico, N. Iqbal, C. Diakhate, and J.-M. Conrat, "On the frequency dependency of radio channel's delay spread: Analyses and findings from mmMAGIC multi-frequency channel sounding," in *12th European Conference on Antennas and Propagation (EuCAP 2018)*, London, United Kingdom (Great Britain), Apr. 2018.

[23] G. L. Stüber, *Principles of Mobile Communication (2Nd Ed.)*. Norwell, MA, USA: Kluwer Academic Publishers, 2001.

[24] N. Iqbal, C. Schneider, J. Luo, D. Dupleich, R. Müller, S. Häfner, and R. S. Thomä, "On the stochastic and deterministic behavior of mmwave channels," in *The 11th European Conference on Antennas and Propagation (EuCAP 2017)*, Mar. 2017.

[25] D. Dupleich, N. Iqbal, C. Schneider, S. Häfner, R. Müller, S. Skoblikov, J. Luo, and R. S. Thomä, "Investigations on fading scaling with bandwidth and directivity at 60 GHz," in *The 11th European Conference on Antennas and Propagation (EuCAP 2017)*, Mar. 2017.

[26] D. Dupleich, N. Iqbal, C. Schneider, S. Häfner, R. Müller, S. Skoblikov, J. Luo, G. Delgaldo, and R. S. Thomä, "Influence of system aspects on fading at mm-waves," *IET Microwaves, Antennas & Propagation*, November 2017. [Online]. Available: http://digital-library.theiet.org/content/journals/10.1049/iet-map.2017.0601

[27] M. Pätzold and B. Talha, "On the statistical properties of sum-of-cisoids-based mobile radio channel models," in *Proc. of WPMC*, 2007, pp. 394–400.

[28] N. Iqbal, J. Luo, C. Schneider, D. Dupleich, R. Müller, S. Häfner, and R. S. Thomä, "Stochastic/Deterministic behavior of cross polarization discrimination in mmwave channels," in *IEEE ICC Wireless Communications Symposium (ICC'17 WCS)*, May 2017.

[29] D. Dupleich, R. Müller, N. Han, S. Häfner, C. Schneider, J. Luo, G. Del Galdo, and R. S. Thomä, "Polarization in spatial channel models at mm-waves- a correlation based approach," in *13th European Conference on Antennas and Propagation (EuCAP)*, 2019, pp. 1–5.

[30] N. Iqbal, J. Luo, R. Müller, G. Steinböck, C. Schneider, D. A. Dupleich, S. Häfner, and R. S. Thomä, "Multipath cluster fading statistics and modeling in millimeter-wave radio channels," *IEEE Transactions on Antennas and Propagation*, vol. 67, no. 4, pp. 2622–2632, April 2019.

[31] N. Iqbal, J. Luo, D. Dupleich, S. Häfner, R. Müller, C. Schneider, and R. S. Thomä, "Second-order statistical characterization of the 60 GHz cluster fading channels," in *2018 IEEE 29th Annual International Symposium on Personal, Indoor and Mobile Radio Communications (PIMRC)*, Sep. 2018, pp. 241–245.

[32] N. Iqbal, D. Dupleich, C. Schneider, J. Luo, R. Müller, S. Häfner, G. Del Galdo, and R. S. Thomä, "Modeling of intra-cluster multipaths for 60 GHz fading channels," in *12th European Conference on Antennas and Propagation (EuCAP 2018)*, Apr. 2018.

[33] M. Pätzold, *Mobile radio channels.* John Wiley & Sons, 2011.

[34] N. Iqbal, J. Luo, C. Schneider, D. A. Dupleich, R. Müller, S. Häfner, and R. S. Thomä, "Investigating validity of wide-sense stationary assumption in millimeter wave radio channels," *IEEE Access*, vol. 7, pp. 180 073–180 082, 2019.

[35] A. F. Molisch, A. Karttunen, R. Wang, C. U. Bas, S. Hur, J. Park, and J. Zhang, "Millimeter-wave channels in urban environments," in *2016 10th European Conference on Antennas and Propagation (EuCAP)*, April 2016, pp. 1–5.

[36] J. Järveläinen, S. L. H. Nguyen, K. Haneda, R. Naderpour, and U. T. Virk, "Evaluation of millimeter-wave line-of-sight probability with point cloud data," *IEEE Wireless Communications Letters*, vol. 5, no. 3, pp. 228–231, June 2016.

[37] T. Rappaport, *Wireless Communications: Principles and Practice*, 2nd ed. Upper Saddle River, NJ, USA: Prentice Hall PTR, 2001.

[38] M. Steinbauer, A. Molisch, and E. Bonek, "The double-directional radio channel," *IEEE Antennas and Propagation Magazine*, vol. 43, no. 4, pp. 51–63, 2001.

[39] C. Oestges and B. Clerckx, *MIMO wireless communications: from real-world propagation to space-time code design.* Academic Press, 2010.

[40] A. Maltsev, A. Pudeyev, A. Lomayev, and I. Bolotin, "Channel modeling in the next generation mmwave Wi-Fi: IEEE 802.11ay standard," in *22th European Wireless Conference*, May 2016, pp. 1–8.

[41] U. G. Schuster and H. Bölcskei, "Ultrawideband channel modeling on the basis of information-theoretic criteria," *IEEE Transactions on Wireless Communications*, vol. 6, no. 7, pp. 2464–2475, July 2007.

[42] M. V. Clark and L. J. Greenstein, "The relationship between fading and bandwidth for multipath channels," *IEEE Transactions on Wireless Communications*, vol. 4, no. 4, pp. 1372–1376, July 2005.

[43] W. Q. Malik, B. Allen, and D. J. Edwards, "Impact of bandwidth on small-scale fade depth," in *IEEE Global Telecommunications Conference*, Nov 2007, pp. 3837–3841.

[44] W. B. Davenport, W. L. Root *et al.*, *An introduction to the theory of random signals and noise*. McGraw-Hill New York, 1958, vol. 159.

[45] R. H. Clarke, "A statistical theory of mobile-radio reception," *The Bell System Technical Journal*, vol. 47, no. 6, pp. 957–1000, July 1968.

[46] G. D. Durgin, T. S. Rappaport, and D. A. de Wolf, "New analytical models and probability density functions for fading in wireless communications," *IEEE Transactions on Communications*, vol. 50, no. 6, pp. 1005–1015, June 2002.

[47] E. Zöchmann, S. Caban, C. F. Mecklenbräuker, S. Pratschner, M. Lerch, S. Schwarz, and M. Rupp, "Better than rician: Modelling millimetre wave channels as two-wave with diffuse power," *arXiv preprint arXiv:1804.03417*, 2018.

[48] J. M. Romero-Jerez, F. J. Lopez-Martinez, J. F. Paris, and A. J. Goldsmith, "The fluctuating two-ray fading model: Statistical characterization and performance analysis," *IEEE Transactions on Wireless Communications*, vol. 16, no. 7, pp. 4420–4432, July 2017.

[49] M. Nakagami, "The m-distribution–A general formula of intensity distribution of rapid fading," in *Statistical methods in radio wave propagation*. Elsevier, 1960, pp. 3–36.

[50] S. O. Rice, "Mathematical Analysis of Random Noise," in *Selected Papers on Noise and Stochastic Processes*, N. Wax, Ed. Dover Pubns, Jun. 1954.

[51] L. J. Greenstein, S. S. Ghassemzadeh, V. Erceg, and D. G. Michelson, "Ricean k -factors in narrow-band fixed wireless channels: Theory, experiments, and statistical models," *IEEE Transactions on Vehicular Technology*, vol. 58, no. 8, pp. 4000–4012, Oct 2009.

[52] M. J. Gans, "A power-spectral theory of propagation in the mobile-radio environment," *IEEE Transactions on Vehicular Technology*, vol. 21, no. 1, pp. 27–38, Feb 1972.

[53] W. C. Jakes and D. C. Cox, Eds., *Microwave Mobile Communications*. Wiley-IEEE Press, 1994.

[54] M. Pätzold, U. Killat, and F. Laue, "A deterministic digital simulation model for suzuki processes with application to a shadowed rayleigh land mobile radio channel," *IEEE Transactions on Vehicular Technology*, vol. 45, no. 2, pp. 318–331, May 1996.

[55] M. Pätzold, U. Killat, F. Laue, and Yingchun Li, "On the statistical properties of deterministic simulation models for mobile fading channels," *IEEE Transactions on Vehicular Technology*, vol. 47, no. 1, pp. 254–269, Feb 1998.

[56] H. Schulze, "Stochastische modelle und digitale simulation von mobilfunkkanälen," *Kleinheubacher Berichte*, vol. 32, 01 1989.

[57] Y. R. Zheng and Chengshan Xiao, "Improved models for the generation of multiple uncorrelated rayleigh fading waveforms," *IEEE Communications Letters*, vol. 6, no. 6, pp. 256–258, June 2002.

[58] M. Peter *et al.*, "Measurement results and final mmMAGIC channel models,," *Deliverable D2*, vol. 2, 2017.

[59] Q. H. Spencer, B. D. Jeffs, M. A. Jensen, and A. L. Swindlehurst, "Modeling the statistical time and angle of arrival characteristics of an indoor multipath channel," *IEEE Journal on Selected Areas in Communications*, pp. 347–360, 2000.

[60] M. Pätzold, B. O. Hogstad, and D. Kim, "A new design concept for high-performance fading channel simulators using set partitioning," *Wireless Personal Communications*, vol. 40, no. 3, pp. 267–279, Feb 2007. [Online]. Available: https://doi.org/10.1007/s11277-006-9189-4

[61] P. Almers, E. Bonek, A. Burr, N. Czink, M. Debbah, V. Degli-Esposti, H. Hofstetter, P. Kyösti, D. Laurenson, G. Matz, A. Molisch, C. Oestges, and H. Özcelik, "Survey of channel and radio propagation

models for wireless mimo systems," *EURASIP Journal on Wireless Communications and Networking*, vol. 2007, no. 1, p. 019070, Feb 2007. [Online]. Available: https://doi.org/10.1155/2007/19070

[62] J. P. Kermoal, L. Schumacher, K. I. Pedersen, P. E. Mogensen, and F. Frederiksen, "A stochastic mimo radio channel model with experimental validation," *IEEE Journal on Selected Areas in Communications*, vol. 20, no. 6, pp. 1211–1226, Aug 2002.

[63] W. Weichselberger, M. Herdin, H. Ozcelik, and E. Bonek, "A stochastic mimo channel model with joint correlation of both link ends," *IEEE Transactions on Wireless Communications*, vol. 5, no. 1, pp. 90–100, Jan 2006.

[64] G. Calcev, D. Chizhik, B. Goransson, S. Howard, H. Huang, A. Kogiantis, A. Molisch, A. Moustakas, D. Reed, and H. Xu, "A wideband spatial channel model for system-wide simulations," *IEEE Transactions on Vehicular Technology*, vol. 56, no. 2, pp. 389–403, March 2007.

[65] E. Damosso, L. Correia, and E. C. D. I. "Telecommunications, Information Market, and Exploitation of Research.", *COST Action 231: Digital Mobile Radio Towards Future Generation Systems : Final Report*, ser. EUR (Series). European Commission, 1999. [Online]. Available: https://books.google.co.uk/books?id=setUHQAACAAJ

[66] R. Buehrer, S. Arunachalam, K. Wu, and A. Tonello, "Spatial channel model and measurements for imt-2000 systems," in *Vehicular Technology Conference, 2001. VTC 2001 Spring. IEEE VTS 53rd*, vol. 1, 2001, pp. 342–346 vol.1.

[67] K. Haneda, J. Poutanen, L. Liu, C. Oestges, F. Tufvesson, and P. Vainikainen, "Comparison of delay and angular spreads between channel measurements and the cost2100 channel model," in *Loughborough Antennas Propagation Conference*, Nov 2010, pp. 477–480.

[68] M. Zhu, G. Eriksson, and F. Tufvesson, "The cost 2100 channel model: Parameterization and validation based on outdoor mimo measurements at 300 mhz," *IEEE Transactions on Wireless Communications*, vol. 12, no. 2, pp. 888–897, February 2013.

[69] M. Shafi, A. F. Molisch, P. J. Smith, T. Haustein, P. Zhu, P. D. Silva, F. Tufvesson, A. Benjebbour, and G. Wunder, "5G: A tutorial overview of standards, trials, challenges, deployment, and practice," *IEEE Journal on Selected Areas in Communications*, vol. 35, no. 6, pp. 1201–1221, June 2017.

[70] S. Jaeckel, "Quasi-deterministic channel modeling and experimental validation in cooperative and massive mimo deployment topologies," Ph.D. dissertation, Universitätsbibliothek TU Ilmenau, 2017.

[71] A. Saleh and R. Valenzuela, "A statistical model for indoor multipath propagation," *IEEE Journal on Selected Areas in Communications*, vol. 5, no. 2, pp. 128–137, February 1987.

[72] M. K. Samimi and T. S. Rappaport, "28 GHz millimeter-wave ultrawideband small-scale fading models in wireless channels," *CoRR*, vol. abs/1511.06938, 2015. [Online]. Available: http://arxiv.org/abs/1511.06938

[73] A. Maltsev, R. Maslennikov, A. Sevastyanov, A. Lomayev, and A. Khoryaev, "Statistical channel model for 60 ghz wlan systems in conference room environment," in *Proceedings of the Fourth European Conference on Antennas and Propagation (EuCAP)*, April 2010, pp. 1–5.

[74] A. Maltsev, A. Pudeyev, I. Karls, I. Bolotin, G. Morozov, R. Weiler, M. Peter, and W. Keusgen, "Quasi-deterministic approach to mmwave channel modeling in a non-stationary environment," in *Globecom Workshops (GC Wkshps), 2014*, Dec 2014, pp. 966–971.

[75] Jose F. Monserrat, Saúl Inca, Jordi Calabuig, and David Martín-Sacristán, "Map-based channel model for urban macrocell propagation scenarios," *International Journal of Antennas and Propagation*, vol. 2015, no. 172501, p. 5, 2015.

[76] ITU, "Guidelines for evaluation of radio interface technologies for IMT-Advanced," ITU Publications, Tech. Rep. ITU-R M.2135, 2008.

[77] R. Müller, R. Herrmann, D. A. Dupleich, C. Schneider, and R. S. Thomä, "Ultrawideband multichannel sounding for mm-wave," in *The 8th European Conference on Antennas and Propagation (EuCAP 2014)*, April 2014, pp. 817–821.

[78] R. Davies, M. Bensebti, M. A. Beach, and J. P. McGeehan, "Wireless propagation measurements in indoor multipath environments at 1.7 ghz and 60 ghz for small cell systems," in *41st IEEE Vehicular Technology Conference, Gateway to the Future Technology in Motion*, May 1991, pp. 589–593.

[79] H. Yang, P. F. M. Smulders, and M. H. A. J. Herben, "Indoor channel measurements and analysis in the frequency bands 2 ghz and 60 ghz,"

in *IEEE 16th International Symposium on Personal, Indoor and Mobile Radio Communications*, vol. 1, Sept 2005, pp. 579–583.

[80] O. H. Koymen, A. Partyka, S. Subramanian, and J. Li, "Indoor mm-wave channel measurements: Comparative study of 2.9 ghz and 29 ghz," in *IEEE Global Communications Conference (GLOBECOM)*, Dec 2015, pp. 1–6.

[81] G. R. MacCartney, T. S. Rappaport, S. Sun, and S. Deng, "Indoor office wideband millimeter-wave propagation measurements and channel models at 28 and 73 ghz for ultra-dense 5g wireless networks," *IEEE Access*, vol. 3, pp. 2388–2424, 2015.

[82] G. R. MacCartney, M. K. Samimi, and T. S. Rappaport, "Exploiting directionality for millimeter-wave wireless system improvement," *CoRR*, vol. abs/1503.05265, 2015. [Online]. Available: http://arxiv.org/abs/1503.05265

[83] T. S. Rappaport, E. Ben-Dor, J. N. Murdock, and Y. Qiao, "38 ghz and 60 ghz angle-dependent propagation for cellular amp; peer-to-peer wireless communications," in *IEEE International Conference on Communications (ICC)*, June 2012, pp. 4568–4573.

[84] R. J. Weiler, M. Peter, T. Kühne, M. Wisotzki, and W. Keusgen, "Simultaneous millimeter-wave multi-band channel sounding in an urban access scenario," in *9th European Conference on Antennas and Propagation (EuCAP)*, May 2015, pp. 1–5.

[85] H. Droste and G. Kadel, "Measurement and analysis of wide band indoor propagation characteristics at 17 ghz and 60 ghz," in *Ninth International Conference on Antennas and Propagation, (Conf. Publ. No. 407)*, vol. 2, Apr 1995, pp. 288–291 vol.2.

[86] T. Jamsä, V. Hovinen, A. Karjalainen, and J. Iinatti, "Frequency dependency of delay spread and path loss in indoor ultra-wideband channels," in *The Institution of Engineering and Technology Seminar on Ultra Wideband Systems, Technologies and Applications*, April 2006, pp. 254–258.

[87] D. M. J. Devasirvathan, "Multi-frequency propagation measurements and models in a large metropolitan commercial building for personal communications," in *IEEE International Symposium on Personal, Indoor and Mobile Radio Communications*, Sep 1991, pp. 98–103.

[88] K. Haneda, J. Järveläinen, A. Karttunen, M. Kyrö, and J. Putkonen, "Indoor short-range radio propagation measurements at 60 and 70 ghz," in *The 8th European Conference on Antennas and Propagation (EuCAP 2014)*, April 2014, pp. 634–638.

[89] S. Häfner, D. A. Dupleich, R. Müller, J. Luo, E. Schulz, C. Schneider, R. S. Thomä, X. Lu, and T. Wang, "Characterisation of channel measurements at 70ghz in indoor femtocells," in *IEEE 81st Vehicular Technology Conference (VTC Spring)*, May 2015, pp. 1–5.

[90] A. Algans, K. I. Pedersen, and P. E. Mogensen, "Experimental analysis of the joint statistical properties of azimuth spread, delay spread, and shadow fading," *IEEE Journal on Selected Areas in Communications*, vol. 20, no. 3, pp. 523–531, Apr 2002.

[91] L. M. Correia, Ed., *Wireless Flexible Personalized Communications - COST 259: European Co-operation in Mobile Radio Research*. John Wiley & Sons, 2001, COST Action 259—Final Report. [Online]. Available: http://www.wiley.com/cda/product/0,,047149836X,00.html

[92] W. Zhang, T. D. Abhayapala, and J. Zhang, "UWB spatia - frequency channel characterization," in *IEEE 63rd Vehicular Technology Conference*, vol. 6, May 2006, pp. 2732–2736.

[93] T. M. Cover and J. A. Thomas, *Elements of information theory*. John Wiley & Sons, 2012.

[94] R. Müller, S. Häfner, D. Dupleich, R. S. Thomä, G. Steinböck, J. Luo, E. Schulz, X. Lu, and G. Wang, "Simultaneous multi-band channel sounding at mm-wave frequencies," in *2016 10th European Conference on Antennas and Propagation (EuCAP)*, April 2016, pp. 1–5.

[95] A. Böttcher, C. Schneider, P. Vary, and R. S. Thomä, "Dependency of the power and delay domain parameters on antenna height and distance in urban macro cell," in *Proceedings of the 5th European Conference on Antennas and Propagation (EUCAP)*, April 2011, pp. 1395–1399.

[96] A. F. Molisch, *Wireless Communications*, 2nd ed. Wiley Publishing, 2011.

[97] M. Zhu, "Geometry-based radio channel characterization and modeling: Parameterization, implementation and validation," Ph.D. dissertation, Lund University, 2014.

[98] J. D. Kraus, D. A. Fleisch, and S. H. Russ, *Electromagnetics*, 5th ed., ser. McGraw-Hill series in electrical and computer engineering). Boston: McGraw Hill, 1999.

[99] M. Shafi, M. Zhang, A. L. Moustakas, P. J. Smith, A. F. Molisch, F. Tufvesson, and S. H. Simon, "Polarized MIMO channels in 3-D: models, measurements and mutual information," *IEEE Journal on Selected Areas in Communications*, vol. 24, no. 3, pp. 514–527, March 2006.

[100] X. Gao, L. Dai, Y. Zhang, T. Xie, X. Dai, and Z. Wang, "Fast channel tracking for terahertz beamspace massive mimo systems," *IEEE Transactions on Vehicular Technology*, vol. 66, no. 7, pp. 5689–5696, July 2017.

[101] B. Picinbono and P. Bondon, "Second-order statistics of complex signals," *IEEE Transactions on Signal Processing*, vol. 45, no. 2, pp. 411–420, Feb 1997.

[102] P. J. Schreier and L. L. Scharf, "Second-order analysis of improper complex random vectors and processes," *IEEE Transactions on Signal Processing*, vol. 51, no. 3, pp. 714–725, March 2003.

[103] F. D. Neeser and J. L. Massey, "Proper complex random processes with applications to information theory," *IEEE Transactions on Information Theory*, vol. 39, no. 4, pp. 1293–1302, Jul 1993.

[104] P. J. Schreier and L. L. Scharf, *Statistical signal processing of complex-valued data: the theory of improper and noncircular signals.* Cambridge University Press, 2010.

[105] A. Hughes, S. Y. Jun, C. Gentile, D. Caudill, J. Chuang, J. Senic, and D. G. Michelson, "Measuring the impact of beamwidth on the correlation distance of 60 ghz indoor and outdoor channels," *IEEE Open Journal of Vehicular Technology*, 2021.

[106] D. A. Dickey and W. A. Fuller, "Distribution of the estimators for autoregressive time series with a unit root," *Journal of the American statistical association*, vol. 74, no. 366a, pp. 427–431, Jun. 1979.

[107] A. F. Molisch, "Ultra-wide-band propagation channels," *Proceedings of the IEEE*, vol. 97, no. 2, pp. 353–371, Feb 2009.

[108] L. Franks, "Carrier and Bit Synchronization in Data Communication - A Tutorial Review," *IEEE Transactions on Communications*, vol. 28, no. 8, pp. 1107–1121, August 1980.

[109] I. Kochanska, "Testing the wide-sense stationarity of bandpass signals for underwater acoustic communications," in *2017 IEEE International Conference on INnovations in Intelligent SysTems and Applications (INISTA)*, Jul. 2017, pp. 484–489.

[110] G. Rafiq and M. Pätzold, "Statistical properties of the capacity of multipath fading channels," in *IEEE 20th International Symposium on Personal, Indoor and Mobile Radio Communications*, 2009, pp. 1103–1107.

List of Figures

List of Tables